Radioisotopes in
Weed Research

Radioisotopes in Weed Research

Edited by Kassio Ferreira Mendes
Federal University of Viçosa

CRC Press
Taylor & Francis Group
Boca Raton London New York

CRC Press is an imprint of the
Taylor & Francis Group, an **informa** business

First edition published 2021
by CRC Press
6000 Broken Sound Parkway NW, Suite 300, Boca Raton, FL 33487-2742
and by CRC Press
2 Park Square, Milton Park, Abingdon, Oxon, OX14 4RN

Library of Congress Cataloging-in-Publication Data

Names: Mendes, Kassio Ferreira, editor.
Title: Radioisotopes in weed research / Kassio Ferreira Mendes.
Description: First edition. | Boca Raton, FL : CRC Press, 2021. | Includes bibliographical references and index. | Summary: "Weed control is a global problem in agriculture, there is constant effort both to develop and utilize herbicides. As part of the general widespread concern over the residual effects of chemicals applied to crop plants, the study of herbicide residues in plants and soil, and the detoxification of herbicides, has become essential. Isotopic techniques can be used to identify degradation products and trace the fate of the herbicide. This book provides: Comprehensive information on the analysis of data collected from research using radioisotopes. Discusses use of radioisotopes of weed research focussing on reviewing absorption, translocation, metabolism, bioaccumulation of pesticides"-- Provided by publisher.
Identifiers: LCCN 2020033896 (print) | LCCN 2020033897 (ebook) | ISBN 9780367643409 (paperback) | ISBN 9780367436612 (hardback) | ISBN 9781003005070 (ebook)
Subjects: LCSH: Radioisotopes. | Herbicides. | Weeds--Research.
Classification: LCC QD466 .R335 2021 (print) | LCC QD466 (ebook) | DDC 628.9/7015413884--dc23
LC record available at https://lccn.loc.gov/2020033896
LC ebook record available at https://lccn.loc.gov/2020033897

ISBN: 9780367436612 (hbk)
ISBN: 9780367643409 (pbk)
ISBN: 9781003005070 (ebk)

Typeset in Palatino LT Std
by KnowledgeWorks Global Ltd.

Contents

Editor

Kassio Ferreira Mendes is Professor of Biology and Integrated Management of Weed, Department of Agronomy, Federal University of Viçosa, Brazil. Post-Doctor (2019) and Doctor in Science (2017), Nuclear Energy in Agriculture (Chemistry in Agriculture and Environment) by the Center of Nuclear Energy in Agriculture, University of São Paulo, Brazil with interuniversity exchange doctorate by University of Minnesota, USA (2016), Twin Cities Campus, College of Food and Agricultural Sciences in the Department of Soil, Water, and Climate. Master in Agronomy (Crop Science) by Federal University of Viçosa (2013). Agronomist by University of State Mato Grosso (2011). Member of Brazilian Society of Weed Science.

Editor

Kassio Ferreira Mendes is Professor of Biology and Integrated Management of Weed, Department of Agronomy, Federal University of Viçosa, Brazil. Post-Doctor (2017), and Doctor in Science (2017), Nuclear Energy in Agriculture (Chemistry in Agriculture and Environment) by the Center of Nuclear Energy in Agriculture, University of São Paulo, Brazil, with an internship/exchange doctorate by University of Minnesota, USA (2016), Twin Cities Campus, College of Food and Agricultural Sciences in the Department of Soil, Water, and Climate. Master in Agronomy (Crop Science) by Federal University of Viçosa (2012), Agronomist by University of Mato Grosso (2011). Member of Brazilian Society of Weed Science.

Contributors

Adijailton Jose de Souza
Department of Soil Science
"Luiz de Queiroz" College
  of Agriculture
University of São Paulo
Piracicaba, Brazil

Alessandro Rizzo
Laboratorio di Sorveglianza
  Ambientale
Istituto di Radioprotezione
Ente Nazionale per le Nuove
  Tecnologie
l'Energia e lo Sviluppo Sostenibile,
Centro Ricerche ENEA Casaccia
Santa Maria di Galeria (RM), Italy

Antonia María Rojano-Delgado
Department of Agricultural
  Chemistry and Edaphology
University of Córdoba
Córdoba, Spain

Douglas Gomes Viana
Department of Soil Science
"Luiz de Queiroz" College
  of Agriculture
University of São Paulo
Piracicaba, Brazil

Felipe Gimenes Alonso
Laboratory of Ecotoxicology
Center of Nuclear Energy
  in Agriculture
University of São Paulo
Piracicaba, Brazil

Gabriel da Silva Amaral
Department of Chemistry
Federal University of São Carlos
São Carlos, Brazil

Kamila Cabral Mielke
Department of Agronomy
Federal University of Viçosa
Viçosa, Brazil

Kassio Ferreira Mendes
Department of Agronomy
Federal University of Viçosa
Viçosa, Brazil

Leonardo Vilela Junqueira
Laboratory of Ecotoxicology
Center of Nuclear Energy in
  Agriculture
University of São Paulo
Piracicaba, Brazil

Luca Ciciani
Istituto di Radioprotezione
Ente Nazionale per le Nuove
  Tecnologie
l'Energia e lo Sviluppo Sostenibile,
Centro Ricerche ENEA Casaccia
Santa Maria di Galeria (RM), Italy

Lucas Heringer Barcellos Júnior
Department of Agronomy
Federal University of Viçosa
Viçosa, Brazil

Maria Fátima das Graças
Fernandes da Silva
Department of Chemistry
Federal University of São Carlos
São Carlos, Brazil

Matheus Bortolanza Soares
Department of Soil Science
"Luiz de Queiroz" College
  of Agriculture
University of São Paulo
Piracicaba, Brazil

Nicoli Gomes de Moraes
Laboratory of Ecotoxicology
Center of Nuclear Energy
  in Agriculture
University of São Paulo
Piracicaba, Brazil

Rafael De Prado
Department of Agricultural
  Chemistry and Edaphology
University of Cordoba
Córdoba, Spain

Ricardo Alcántara de la Cruz
Department of Chemistry
Federal University of São Carlos
São Carlos, Brazil

Rodrigo Nogueira de Sousa
Department of Soil Science
"Luiz de Queiroz" College
  of Agriculture
University of São Paulo
Piracicaba, Brazil

Te-Ming Tseng
Department of Plant and Soil
  Sciences
Mississippi State University
Starkville, Mississippi, USA

Valdemar Luiz Tornisielo
Laboratory of Ecotoxicology
Center of Nuclear Energy
  in Agriculture
University of São Paulo
Piracicaba, Brazil

Vanessa Takeshita
Laboratory of Ecotoxicology
Center of Nuclear Energy
  in Agriculture
University of São Paulo
Piracicaba, Brazil

Ziming Yue
Department of Plant and Soil
  Sciences
Mississippi State University
Starkville, Mississippi, USA

chapter one

Historical use of radioisotopes in weed research

Ziming Yue and Te-Ming Tseng
Mississippi State University

Contents

1.0 Introduction

An atom consists of a certain number of protons and different numbers of neutrons. The former determines its chemical property, and the latter determines its atomic weight. An atom with an unstable number of protons or neutrons tends to disintegrate to release different particles (such as β-particles) or photons (such as γ-rays). The discovery of radioactivity was attributed to the pioneering work of Röntgen (1895) and Becquerel (1896). Further investigation by Curie and Curie (1898), and Rutherford (1911) established that radioactivity is exhibited by heavy elements such as uranium, thorium, and radium. Large-scale application of radioactivity began at the end of World War II in the Manhattan Project, during which the scintillation counter was invented by Sir Samuel Curran in 1944 (Curran, 1949). The liquid scintillation counter (LSC) was first commercialized by Packard as TriCarb 314 in 1953.

Coincidently, modern synthetic herbicides such as 2,4-D and MCPA were commercialized as "Weedone" by the American Chemical Paint Company in the US in 1945, and as "Agroxone" by Imperial Chemical Industries (ICI) in the UK in 1946, respectively. Their independent discoveries could be traced back to 1941 when Templeman (ICI), and Nutman and collaborators (Rothamsted Experimental Station) first demonstrated the herbicidal activity of MCPA in the UK. Pokorny in 1941 synthesized 2,4-D, while Zimmerman and Hitchcock in 1942 characterized its growth regulatory properties in the US. The herbicide technology, combined with the availability of the radioactivity detection technology, especially LSC, led to the application of radioisotopes in weed research. Such a combination soon contributed to weed physiology and ultimately weed control. The invention of the autoradiography technique to detect radio-isotope translocation in plants was attributed to Crafts (Yamaguchi and Crafts, 1958; Zsweep, 1961). This technique allowed ^{14}C-labeled herbicide or other organic compounds to be applied on the surface of the leaf, stem, or root. After allowing some time for the ^{14}C compound to be absorbed and translocated in the plant, the plant is then heated to 60°C for 10 minutes to kill the plant, and then mounted to press and dry for 1 week (ideally freeze-dried to avoid any movement of the herbicide during drying). The dried plant (part) is then pressed on an x-ray film for 3–6 weeks to collect sufficient radiation for exposure (excessive pressure avoided), after which a film is developed to capture an image. This technique had a low detection limit and allowed the study of herbicide translocation in plants at physiological ranges. Because of the simplicity and reliability of the technique, it soon spread more widely throughout the academic community than LSC in the 1960s and lasted 60 years until now, where it is still widely used.

Another milestone was the symposium on the Use of Isotopes in Weed Research, which was convened jointly by the Food and Agriculture Organization of the United Nations and the International Atomic Energy Agency (IAEA) and was held in Vienna at the Headquarters of IAEA in October 1965. The symposium collected six papers on herbicide absorption and translocation, six papers on herbicide metabolism, and eight papers on practical techniques on the subject, with almost all the topics in the field being covered. Since then, 55 years ago, more progress has been achieved in this field. Recently, Nandula and Vencill (2015) and Mendes et al. (2017) reviewed the general methods, procedure, and equipment for absorption and translocation of ^{14}C-labeled herbicide in weeds excellently. This paper aimed at reviewing broader topics and focusing on the contributions of radioisotopes to weed research by following the historical development in the field. During the selection of references, personal judgment was applied, and complete coverage was not sought. In addition, following the tradition of IAEA (1966) and journal articles in this field, crop species were also covered in the review.

1.1 Traditional radioactive isotope application in weed research

1.1.1 Liquid scintillation counting

1.1.1.1 Equipment

In weed research, the two dominant ways to detect radioactivity were through LSC and autoradiography, although early authors used a Geiger tube (Leonard and Hull, 1966). While the latter is simple and visual, the former requires at least one piece of equipment, which has gone through several generations. The LSC mixes a liquid sample with a cocktail (usually includes a solvent, emulsifier, and scintillator), and the scintillator absorbs the radiation energy and emits light, which is detected by photomultipliers (PMT) (two or three, depending on the equipment) (Figure 1.1). The early PMT was subject to high noise (background) and had to cool down to around 5°C to reduce noise. Later improvements in PMT technology allowed it to work at room temperature. Since the first commercialized LSC by Packard Instrument Company (TriCarb 314) in 1953, many manufacturers entered the market during the 1960s–1970s including the Nuclear Chicago (later Searle-Analytic), USA; Beckman Instruments, USA; Intertechnique, France; Wallac, Finland; Phillips, Holland; and Hitachi AccuFLEX (later Hitachi Aloka), Japan. In the 1980s, the TriCarb

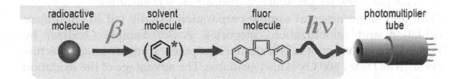

Figure 1.1 Principle of liquid scintillation counting (https://www.perkinelmer .com/lab-products-and- services/application-support-knowledgebase/radiometric/ liquid-scintillation-counting.html).

series introduced time-resolved (TR) counting technique, which significantly lowered background noise (1987). Wallac Oy of Finland introduced Quantulus, which used a guard detection system to reduce background (1985). Perkin Elmer entered the field by acquiring Wallac and Packard Bioscience from the 1990s to 2001, and it kept their fundamental designs unchanged except the software. Hidex, Finland entered the field by launching Hidex Triathler in 1993, which was the first portable LSC with a single PMT. There was no significant improvement in technology since the 1980s until the launch of the Hidex 300 SL in 2008 (Eikenberg et al., 2014). The Hidex 300 SL is a new-generation automatic counter with a different design than the conventional coincidence counter based on two PMTs. It utilized three PMTs aligned at 120° from each other. This detection geometry yielded exceptionally high counting efficiency and counting of samples in triple mode with no risk of luminescence interference from the background. The three-PMT design also enabled triple to double coincidence ratio counting (TDCR), which is an absolute counting method for obtaining the counting efficiency of the samples without external or internal standard sources (Temple, 2015). The TDCR was established as a unique technique by Hidex. The major commercial LS manufacturers that remain at present are PerkinElmer (TriCarb and Quantulus), Hidex, and Aloka.

1.1.1.2   Cocktail

Cocktail or organic scintillator made LSC and alpha counting possible (Passo, 2013). The early cocktail contained benzene, p-dioxane, toluene, and xylene, some of which were miscible with aqueous samples and biological specimens containing methanol, ethylene glycol, and glycol ethers. Later cocktails were based on toluene or xylene mixed with Triton, a non-ionic surfactant, and were also called "modern" cocktails in the 1960s. However, toluene and xylene are both toxic, threatening operators' health, and even now toluene is still used in counting standards such as quench standards (sealed). The first environmentally safe and non-toxic cocktail (Opti-Fluor) appeared in 1984, and the present cocktails are all non-toxic, such as the Ultima Gold (Passo, 2013).

1.1.1.3   Sample preparation: oxidizer vs. solubilization

Plant and environmental sample preparation usually fall in equipment oxidation and solubilization categories. A 307 Sample Oxidizer by PerkinElmer is one option, and similar equipment from other manufacturers such as Hidex 600 Ox is also available. The advantage of the oxidation method is that it is easy to get a colorless and homogeneous solution to avoid color quench and chemical quench (PerkinElmer Application Note), but the equipment investment and carry-over contamination (or recovery) is a concern for this method.

Besides sample oxidation, samples can also be solubilized directly with specific protocols depending on the sample property. Sample solubilization often divides into the following three categories:

1. Alkaline systems (e.g., NaOH)
2. Acidic systems (e.g., $HClO_4$)
3. Other systems (e.g., NaClO)

The most useful reagent is sodium hypochlorite, whose mode of action is via oxidative bleaching. This is particularly useful when dealing with plant samples, especially those containing chlorophyll, where the sodium hypochlorite effectively prevents color quench in subsequent liquid scintillation (LS) counting by bleaching out all of the color present (PerkinElmer Application Note, 2008).

1.1.2   Absorption, translocation, and distribution

As the inventor of autoradiography, Crafts (1966) conducted radioisotope tracer studies using a herbicide such as 2,4-D, and a fertilizer such as urea to determine absorption and translocation primarily through the symplastic movement via the phloem. He was primarily concerned about the physiological level of the herbicide (low concentration range). The bud or shoot apical meristem (SAM) development required not only nutrients from the transpiration stream or apoplastic flow but also the food from the symplastic or phloem flow. The auxin herbicide, 2,4-D, was effective in controlling broadleaf weed species; thus, it was often applied on cereal grains (Kirby, 1980). In terms of barley, the leaf-applied 2,4-D moved to the stem, roots, and SAM at the 1-, 2-, 3-, and 4-leaf stages, but at the 5- and 6-leaf stages it rarely moved out of the applied leaf (Zweep, 1961). The lower damage at a later age by 2,4-D in the field was due to the lack of translocation out of the leaves, rather than the increasing resistance of SAM and roots (Zweep, 1961; Petersen, 1966).

Herbicide translocation, on the other hand, is slightly different in plants with woody vines. Hutchison et al. (2010) studied the translocation of different herbicides (glyphosate, metsulfuron, and triclopyr) in Old World Climbing Fern (OWCF) (*Lygodium microphyllum*), an invasive fern vine in forest wetlands in southern Florida. The fern can grow up to 15 m high to catch the canopy, and a primary problem is its ability to resprout from the rhizome. Hence, the herbicide translocation into the rhizome is the key to achieving effective control of this invasive species. The results showed that the majority of radioactivity remained in the treated leaves for all herbicides, with only a small percentage of the absorbed radioactivity being detected in other plant parts. All three herbicides translocated acropetally and basipetally to some extent. Radioactivity, for the most part, translocated evenly throughout the plants, but the greatest amount of radioactivity derived from triclopyr occurred in the rhizomes, suggesting that triclopyr may be more effective in killing the rhizomes. They suggested a lower band foliar application + cut and spray (cut the rachis to expose phloem to load the herbicides directly into phloem to be transported to rhizomes) for the OWCF control.

García-Angulo et al. (2009) used ^{14}C-labeled glucose to trace the changes in bean cells that habituated a lethal dose of dichlobenil suspension for 3–5 years. It was found that the incorporation of ^{14}C-labeled glucose into cellulose in dehabituated cells was reduced by 40% in the absence of dichlobenil when compared to non-habituated cells. If the unhabituated cells were re-exposed to dichlobenil, then the incorporation of ^{14}C-labeled glucose into cellulose increased by 3.3-fold. Results indicated that the habituation of bean-cultured cells to dichlobenil leads to a stable change in the cellulose biosynthesis complex.

1.1.3   Plant carbon metabolism

Plants fix atmospheric CO_2 as their primary carbon source, and some organic compounds also contributed as a minor carbon source. Rice was reported to absorb 1%–14% of its carbon from roots under flooded conditions. The CO_2 fixation by soybean root nodule reached 30% of the nodule respiration supplying carbon skeletons for amino acid biosynthesis (Coker and Schubert, 1981). Okada and Kumula (1986) grew sweet potato plants in different soil matrix with different ^{14}C concentrations. They used gas proportional counter and mass spectrometer quantified carbon isotope signatures of the sweet potato plant parts and concluded that a significant amount of soil carbon (compost) was assimilated as sweet potato biomass, in both leaves and roots. In addition, Kigoshi et al. (1984) used different ^{14}C-signatured soil CO_2 environments to grow sweet potato; e.g., compost was a high ^{14}C soil and soil under the floor was a low ^{14}C soil, while leaves were in atmospheric CO_2 environment. They found the sweet potato leaves showing the atmospheric CO_2 ^{14}C isotope signature, while the root starch ^{14}C signature was influenced by the soil CO_2 ^{14}C signature. The results suggested there was another CO_2 fixation pathway beyond leaf photosynthesis.

1.1.4   The fate of ^{14}C-labeled herbicide in plants and environments

Slade (1966) was one of the early explorers who investigated paraquat metabolism in crops. He applied ^{14}C-labeled paraquat on potato tops and found the leaves to be dead in a few days. There were only minor photochemical degradation products in the dead leaves (4-carboxy-1-methylpyridinium chloride and methylamine hydrochloride), but the translocated radioactivity in harvested potato tubers was due to the presence of paraquat (level 0.08 ppm).

Herbicides 2,4,5-trichlorophenoxyacetic acid (2,4,5-T) and 2-methyl-4-chlorophenoxyacetic acid (MCPA) are soluble only in diesel oil, so they were often applied in diesel solutions (Suss et al., 1966). Suss et al. (1966) used ^{14}C-labeled pentadecane as one of the main components of diesel to investigate the decomposition of diesel in different soils. The measurements of the ^{14}C showed that the decomposition of diesel depends largely on the activity of the microorganisms in the soil. For soils with lower content of microorganisms, decomposition took longer to begin. After 3 years, about 70% of the ^{14}C activity in the organic matter belonged to humic acid regardless of the two experimental soils.

Lorraine-Colwill et al. (2003) collected fractions after HPLC at different times and used liquid scintillation counter to document glyphosate metabolism at six days after treatment. No metabolite aminomethylphosphonic acid (AMPA) was observed. If the metabolite was not known, the radioactive label was an effective way to locate the metabolite using the label.

Canny and Markus (1960) used Geiger tube (EW3H) measured the evolution of labeled CO_2 from roots and shoots of a tick bean plant treated with ^{14}C labeled 2,4-D on one leafet. The CO_2 evolved from the roots was consistently more radioactive than the from the shoots, suggesting the breakdown rate of 2,4-D was much higher in roots than in shoots in tick bean plants.

1.1.5   Determination of crop domestication time and biofuel analysis

The domestication time of crops was usually determined by ^{14}C dating, which depends on ^{14}C radioactivity measurement using the LSC. The radiocarbon dating was invented by Willard Libby and colleagues in 1949 (Libby, 1955), who was awarded the Nobel Prize in chemistry for this work in 1960. They originally used a modified Geiger counter for analysis, which was soon superseded by the gas proportional counting (GPC) technique. The LSC was first used in radiocarbon dating in the early 1950s, and the recent approach for radiocarbon dating is an accelerator mass spectrometer (AMS), where $^{14}C/^{12}C$ is directly measured. The age of the sample was calculated by the radioactive decay law (Woods Hole Oceanographic Institution, 2007):

$$N(t) = N_0 e^{-0.000120968t}$$

N(t) – Number of ^{14}C atoms at time t (years), referring to the present time.
N_0 – Number of ^{14}C atoms at time 0 (years), referring to the time when those ^{14}C atoms were assimilated or formed the sample.

In radiocarbon dating using LSC, the sample usually needs to be converted into benzene, while in radiocarbon dating by AMS, the sample needs to be converted into graphite. Compared to the LSC method, the AMS method requires a smaller sample, takes less time (usually less than 24 hours), and has higher precision, but is also more expensive (Bronic et al., 2009). In addition, LS counting is also used to analyze biofuel. Since plant-originated biofuel contains ^{14}C from the atmospheric CO_2, the LS counting of ^{14}C radiation of biofuel can get the biofuel concentration, and because it is a simple analytical method, no special user training is required (Krištof and Logar, 2017).

1.2   Recent progress

1.2.1   Root exudate and allelopoathy

Plants are capable of exuding many kinds of endogenous compounds from their roots into the surrounding soil (Linder et al., 1964). These compounds can suppress the growth of adjacent plants, and this phenomenon is called allelopathy. It plays an essential role in plant–plant interaction in agriculture and ecology. Dinelli et al. (2006) studied the translocation and root exudation of herbicide after foliar treatments of wheat and ryegrass using ^{14}C-labeled diclofop-methyl and triasulfuron. The results showed the presence of untreated plants (wheat or ryegrass) in the same pot as triasulfuron-treated ryegrass or wheat induced the exudation of the herbicide 7 to 32 times more. In the case of diclofop-methyl, the induced root exudation of the herbicide was three to six times more in the presence of untreated wheat or ryegrass. The root-exuded herbicides were inhibiting the adjacent plants, so this was a form of allelopathy. The implication is that herbicide root exudation and transfer could be significant in the field.

Dinelli et al. (2006) indicated that glyphosate had a similar influence on the root exudation of ryegrass, i.e., root exudates of glyphosate increased with time under the sub-lethal dose of glyphosate. Yanncarri et al. (2012) showed that root exudation of radiolabeled glucose decreased significantly under the influence of glyphosate in susceptible ryegrass populations. This indicates that the root exudation rates of assimilates, such as glucose and allelopathic compounds, such as glyphosate, were different. This difference may be because the phloem is a living tissue, and the cell membrane may have control over the flow of different compounds.

1.2.2   Application of other radioactive nuclides in weed research

Radionuclides other than ^{14}C as tracers in weed research were explored early in history. Rademachher et al. (1961) labeled seeds of Sinapis arvensis (wild mustard) and *Avena fatua* (wild oat) with ^{35}S by root assimilation. They found that the length of dormancy is substantially influenced by the manner of soil cultivation. As the half-life of ^{35}S (87.1 days) was too short compared to the dormancy of more than a year, Kloke and Riebartsch (1966) explored the possibility of rare earth elements as tracers for seed dormancy. They labeled weed plants of *Vicia villosa* (hairy vetch) and *Sinapsis alba* (white mustard) by feeding them with fertilizers. The plant roots, stems, leaves, and seed were harvested and irradiated in the atomic reactor, and the rare earth elements were determined by neutron-activation analysis. The results showed that traceable quantities of lanthanum, europium, and dysprosium were present in roots, stems, and leaves, but not in the seed grains. Possibly, a greater supply of these elements might increase their incorporation in the seeds. If so, europium and dysprosium were suitable tracers, while lanthanum had 100 times higher background. Such tracers could possibly be used to study dormancy for several years.

Arsenicals represents a family of herbicides whose use is limited in modern agriculture due to environmental and food security concerns. Arsenical herbicides are, at present, only used in cotton production. Early weed scientists had used ^{74}As-labeled sodium arsenite (8 ppm) to study its effects on pond ecology (Ball et al., 1966). The arsenic label was traced through the food webs, water, and sediments of the pond and aquaria. It was found that ^{74}As moved steadily into the sediments, and the herbicide killed all higher aquatic plants. Plants uptake arsenic rapidly and it stays in the dead plants during their decay. The herbicide decreased pond metabolism and caused an immediate reduction in the rate of carbon fixation by phytoplankton. The reduced rate lasted 10 days. Certain groups of invertebrates disappeared, others decreased in numbers, and recovery of invertebrates was lagging. All fishes were still alive but decreased in their activity.

Gilgen and Feller (2013) used ^{57}Co and ^{65}Zn to trace the solute allocation in broadleaf dock (*Rumex obtusifolius*) under drought stress. They found that the solute moved to the roots under drought conditions and was reversible upon rewatering, which explained their adaptability under drought conditions. This is important because with climate change and freshwater resource exhaustion, drought condition will impact both crops and weeds and their competition. The success of this research was based on the known mobility of these heavy metals in the phloem (Page and Feller, 2005; Riesen and Feller, 2005). The radioactivity of ^{57}Co and ^{65}Zn was determined by a gamma counter (1480 Wizard 30, Wallac, Turku, Finland).

1.2.3 Glyphosate resistance

Glyphosate is the most widely used herbicide in the world. Since it was discovered in 1974, it has at least two significant contributions to agriculture:

no-till farming in the 1980s and glyphosate-resistant crops in the 1990s (Powles and Preston, 2006; Koning et al., 2019). Glyphosate-resistant weeds were first reported 20 years after glyphosate was commercialized. At present, there are eight species reported to be glyphosate resistance. Confirmed resistance mechanisms include mutation(s) of the EPSPs gene, enhanced EPSPs gene copy number, and self-limiting translocation. Radioactive isotopes provided evidence for self-limiting translocation (Wakelin et al., 2004). The leaf-applied glyphosate in the resistant *Lolium rigid* moved to the applied leaf tip. In contrast, the leaf-applied glyphosate in the susceptible *L. rigid* moved to the stem meristematic portion of the plant twice as much compared to the resistant plant.

While Crafts (1966) used physiological-level ^{14}C-labeled 2, 4-D and autoradiography to study the symplastic transport in plants, Yannicari et al. (2012) more recently applied ^{14}C-labeled glucose on the leaf to study differential translocation between a susceptible and resistant population of *Lolium perenne* L. in the presence and absence of glyphosate. The results showed that glucose absorption did not vary under glyphosate treatment, regardless of the susceptible or resistant populations. However, after absorption of the labeled glucose, the translocation of ^{14}C label and its distribution patterns were significantly affected by glyphosate in one day for the susceptible population. The lower ^{14}C translocation significantly affected the unexpanded leaves and the apical meristem on the labeled tiller. In addition, the ^{14}C exudates from roots were decreased considerably in the presence of glyphosate, but only in the susceptible plants. This is in contrast to the resistant population, where glyphosate did not influence leaf absorption, translocation, or root exudation of ^{14}C-labeled glucose and metabolites.

1.2.4   Progress in techniques

1.2.4.1   Autoradiographic imaging
Crafts invented the autoradiography technique in 1958 and it has been used in radioactive isotope research in weed science as a classic technique (Yamaguchi and Crafts, 1958; Zweep, 1961). The x-ray film exposure was used to capture the autoradiography, and it usually took 3–6 weeks to get sufficient exposure (Yamaguchi and Crafts, 1958; Peterson 1966). Now, with phosphorescence imaging (PI), the exposure time is shortened to a day without the degeneration of image quality (Figure 1.2 and Figure 1.3) (Wehtje et al., 2007). This may eliminate several weeks of waiting to get enough exposure, while the herbicide translocation in plants only takes 5 hours. In this technique, a freeze dryer is often used to stop experiments on time to avoid further movement of the herbicide in vascular tissue during drying, while sometimes heat (60°C) is used to kill the plant to stop the experiment.

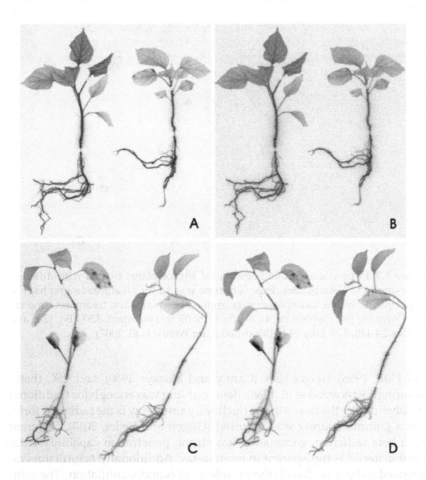

Figure 1.2 Autoradiographs of smallflower morningglory treated with root-applied ^{14}C-atrazine. Foliar radiation dosage in the mount used in A and B was 67.57 Bq/mg. (A) Five x-ray film, no plastic wrap, and 3-wk exposure; (B) five PI, with wrap and 3-d exposure. Dosage in (C) and (D) was 22.87 Bq/mg. (C) Five x-ray film with no plastic wrap and 3-wk exposure; (D) five phosphorescence imaging with plastic wrap (PPI), and 3-d exposure. (Modified from Wehtje et al. 2007.)

1.2.4.2   Heavy metal radionuclides and gamma counting

Traditional labeling was ^{14}C labeling, usually on herbicide, while the recent label uses ^{14}C labeling on glucose to trace photosynthate flow (Yanniccari et al., 2012). Moreover, with the accumulated understanding of heavy metal mobility in the phloem (Page and Feller, 2005; Riesen and Feller, 2005), heavy metal radionuclides were used to trace flow in the phloem.

For the quantitation of the radioactivity in weed research, early researchers used blackness of the x-ray film in autoradiography (Leonard

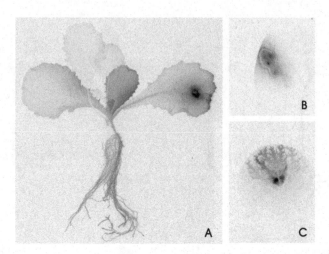

Figure 1.3 PI-produced autoradiographs of lettuce plants treated with different foliar-applied ^{14}C-herbicides. Exposure time was 1d for all autoradiographs. The respective herbicide, radiation dosage applied per plant, and treatment time are as follows: (A) five glyphosate, 44.10 kBq, 60 h; (B) five paraquat, 8.33 kBq, 12 h; and (C) five 2,4-DB, 8.75 kBq, 24 h. (Modified from Wehtje et al. 2007.)

and Hull, 1966), Geiger tube (Canny and Markus 1960), and LSC (beta-counting) (Kirkwood et al. 1966). Beta counting was enough for traditional ^{14}C labeling; for the heavy metal nuclides, gamma-ray is the radiation form, and a gamma counter was required (Gilgen and Feller, 2013). Different from beta radiation, gamma-ray has strong penetration capability since plant material is transparent to gamma-ray. Additionally, scintillators do not need to be mixed with the sample as in liquid scintillation. The scintillator (usually NaI crystal) surrounding the sample absorbs gamma-ray from the sample and emits visible photon for PMT detection. The dried plant parts and environmental samples can be counted directly, as Gilgen and Feller (2013) did. Plastic sample vials are recommended over glass vials, which contain ^{40}K, a radiation source, which may increase background. For the ^{14}C beta decay, the maximum beta particle kinetic energy is 156 keV, but the average weighted energy is 49 keV; for the ^{65}Zn decay, the major gamma-ray has an energy of 1.116 MeV and is emitted in 51% of the decays (Furberg, 1951); for the ^{57}Co decay, the major gamma-ray is emitted at 122 keV and 136 keV (Ricci et al., 1960).

1.2.4.3 Oxidizer-free liquid scintillation

The detection of heavy metal radionuclides such as ^{65}Zn and ^{57}Co requires a gamma counter (NaI crystal scintillator). As gamma-ray has much stronger penetrating power, samples do not need to be oxidized, but sampled as they are (Gilgen and Feller, 2013). Even for ^{14}C LSC, sample solubilization without an oxidizer is also an option. Yannicari et al. (2007) used 9N NaOH solution to digest the dried plant part sample and homogenized before taking an aliquot for mixing with the scintillant and counting. It is believed that acceptable results were obtained, but 9N NaOH is unable to destroy chlorophyll molecules; thus, the green color of the sample may lead to a color quench in the liquid scintillation (PerkinElmer, Application note, 2008). In addition, the samples of Yannicari et al. (2007) were not completely dissolved due to cellulose, while sample homogeneity is usually required for liquid scintillation counting (PerkinElmer, Application note, 2008). Hence, sample solubilization is not a standard protocol, and the goal is to get acceptable results depending on the sample or project. It generally takes a long time or more steps to confirm that there is no risk of color or chemical quench or sample homogeneity risk.

1.3   Concluding remarks

Radioactive isotopes have been used in weed research for more than 70 years. Due to their uniqueness, they have provided many indispensable contributions to weed physiology and weed control. The ^{14}C is the most commonly used isotope in this field. At the same time, radioactive isotope detection technology has also progressed significantly. Modern liquid scintillation technology has recognized and removed backgrounds due to ^{40}K in glass counting vials in addition to low-noise PMT, TR, and TDCR technologies, sample preparation to benzene, and efficient and safer cocktails. AMS is not a direct radiation detection technology; it can provide isotope ratios $^{14}C/^{12}C$ and $^{13}C/^{12}C$ directly, thus offering isotope fractionation correction and double isotope data support (Okada and Kumura, 1986). These technologies have provided numerous possibilities for our experimental design in weed research. For example, weed scientists used to spike ^{14}C-labeled compounds on plants to trace their absorption, translocation, and metabolism and fate in the environment. Their detection is above the plant matter background originated from atmospheric CO_2 that contains ^{14}C. Modern technology has provided the capability to measure the radiation level below this background. If experiments are designed with radiation level on or below plant matter background, fundamental contributions can still be achieved; e.g., Okada and Kumura (1986) demonstrated soil carbon can be a significant part of sweet potato carbon in leaf and roots, and Kigoshi et al. (1984) further demonstrated sweet potato leaf photosynthesis was not the only carbon source for its root starch. An advantage of such low ^{14}C-level research is its ease of disposing radioactive waste and passing regulatory control, which was one of the focuses of Nandula and Vencill (2015). Hence, the appropriate use of modern ^{14}C technology will make a more significant contribution in weed research. Another direction is the application of other radionuclides in weed research. Similar to further implementation of ^{14}C in weed research, based on the possibilities that new technologies have provided, the use of other radionuclides such as heavy metals is based on accumulated research on mobilities of these radionuclides in phloem and gamma counting technology.

References

Ball R.C. and Hooper F.H. 1966. Use of ^{74}As-tagged sodium arsenite in a study of effects of a herbicide on pond ecology. In *Isotopes in Weed Research Proceedings of the Symposium on the Use of Isotopes in Weed Research*, ed. International Atomic Energy Agency, 149–161. IAEA, Vienna, Austria.

Becquerel A.H. 1896. On the rays emitted by phosphorescence. *Comptes Rendus* 122:420–421.

Bronic I.K., Horvatincic N., Baresic J. and Obelic B. 2009. Measurement of 14C activity by liquid scintillation counting. *Applied Radiation and Isotopes*. 67: 800–804.

Canny MJ, Markus K. 1960. The breakdown of 2,4-dichlorophenoxyacetic acid in shoots and roots. *Australian Journal of Biological Sciences*. 13: 486–500.

Coker G.T. and K.R. Schubert. 1981. Carbon dioxide fixation in soybean roots and nodules. *Plant Physiology* 67:691–696.

Crafts AS. 1966. Relation between food and herbicide transport. In *Isotopes in Weed Research Proceedings of the Symposium on the Use of Isotopes in Weed Research*, ed. International Atomic Energy Agency, 3–6. IAEA, Vienna, Austria.

Curie P. and Curie M.S. 1898. On a new radioactive substance contained in pitch-blende. *Comptes Rendus* 127:175–178.

Curran S.C. 1949. *Counting Tubes, Theory and Applications*. Academic Press, New York. p.235.

Dinelli G., Bonetti A., Marotti I., Minelli M., Busi S. and Catizone P. 2006. Root exudation of diclofop-methyl and triasulfuron from foliar-treated durum wheat and ryegrass. *Weed Research* 47:25–33.

Eikenberg J, Beer H, Jäggi M. 2014. Determination of ^{210}Pb and ^{226}Ra/^{228}Ra in continental water using HIDEX 300SL LS-Spectrometer with TDCR efficiency tracing and optimized α/β-discrimination. *Applied Radiation and Isotopes*. 93: 64–69.

Franz JE, Mao MK, Sikorski JA. 1997. Glyphosate: A unique and global herbicide. *ACS Monograph No. 189*. American Chemical Society, Washington.

Furberg S. 1951. The decay scheme of zinc-65. *Nature*. 168: 1005–1006.

García-Angulo P, Alonso-Simón A, Mélida H, Encina A, Acebes JL, Álvarez JM. 2009. High peroxidase activity and stable changes in the cell wall are related to dichlobenil tolerance. *Journal of Plant Physiology*. 166: 1229–1240.

Gilgen AK, Feller U. 2013. Drought stress alters solute allocation in broadleaf dock (*Rumex obtusifolius*). *Weed Science*. 61: 104–108.

Hutchinson J, Langeland K, MacDonald G, Querns R. 2010. Absorption and translocation of glyphosate, metsulfuron, and triclopyr in old world climbing fern (*Lygodium microphyllum*). *Weed Science*. 58: 118–125.

International Atomic Energy Agency (IAEA). 1966. *Isotopes in Weed Research Proceedings of the Symposium on the Use of Isotopes in Weed Research*, ed. IAEA, Vienna, Austria.

Jadiyappa S. 2018. Radioisotope: Applications, effects, and occupational protection. In *Principles and Applications in Nuclear Engineering - Radiation Effects, Thermal Hydraulics, Radionuclide Migration in the Environment*, eds. ROA Rahman and HE-DM Saleh. IntechOpen.

Kessler M. 1990. Time-resolved liquid scintillation counting. *Radiocarbon*. 32: 381–386.

Kessler MJ. 1988. Recent advances in detectors for radioisotopes. In *Research Instrumentation for the 21st Century*, ed. GR Beecher, 1–20. Beltsville Symposia in Agricultural Research, vol 11. Springer, Dordrecht.

Kigoshi K, Kakiuchi N, Shiraki M. 1984. Utilization of carbon dioxide in soil by plant assimilation. In *Report of the International Conference on Nuclear and Radiochemistry*. Lindau, Germany.

Kloke A, Riebartsch K. 1966. Weed seed labelling with rare earth elements. In *Isotopes in Weed Research Proceedings of the Symposium on the Use of Isotopes in Weed Research*, ed. International Atomic Energy Agency, 165–170. IAEA, Vienna, Austria.

Koning LA, de Mol F, Gerowitt B. 2019. Effects of management by glyphosate or tillage on the weed vegetation in a field experiment. *Soil Tillage Res*. 186: 79–86.

Krištof R, Logar JK. 2017. Liquid scintillation spectrometry as a tool of biofuel quantification. In *Frontiers in Bioenergy and Biofuel*, eds. E Jacob-Lopes and LQ Zepka, 59–69. IntechOpen.

Leonard OA, Hull RJ. 1966. Translocation of ^{14}C-labelled substance and ^{32}PO$_4$ in mistletoe-infected and uninfected conifers and dicotyledonous trees. In *Isotopes in Weed Research Proceedings of the Symposium on the Use of Isotopes in Weed Research*, ed. International Atomic Energy Agency, 31–45. IAEA, Vienna, Austria.

Libby WF. 1955. *Radiocarbon dating*, 2d ed., University of Chicago Press.

Linder PJ, Mitchell JW, Freeman GD. 1964. Persistence and translocation of exogenous regulating compounds that exude from roots. *Journal of Agricultural and Food Chemistry*. 12: 437–438.

Lorraine-Colwill DF, Powles SB, Hawkes TR, Hollinshead PH, Warner SAJ, Preston C. 2003. Investigations into the mechanism of glyphosate resistance in Lolium rigidum. *Pesticide Biochemistry and Physiology*. 74: 62–72.

McWhorter C, Jordan T, Wills G. 1980. Translocation of ^{14}C-glyphosate in soybeans (*Glycine max*) and Johnsongrass (*Sorghum halepense*). *Weed Sci*. 28: 113–118.

Mendes KF, Martins BAB, Reis FC, Dias ACR, Tornisielo VL. 2017. Methodologies to study the behavior of herbicides on plants and the soil using radioisotopes. *Planta Daninha*. 35: e017154232

Nandula VK, Vencill WK. 2015. Herbicide absorption and translocation in plants using radioisotopes. *Weed Science*. 63: 140–151.

Okada K, Kumula A. 1986. Uptake of organic matter by roots of sweet potato: Analysis of δ^{14}C value of plants. *Plant and Soil*. 91: 209–219.

Page V, Feller U. 2005. Selective transport of zinc, manganese, nickel, cobalt and cadmium in the root system and transfer to the leaves in young wheat plants. *Annals of Botany*. 96: 425–434.

Passo C. 2013. The evolution of liquid scintillation technique: A personal perspective. Presented at: LSC 2013. Barcelona, Spain, 17–22.

PerkinElmer, Application note: LSC in practice: LSC sample preparation by solubilization. https://www.perkinelmer.com/lab-solutions/resources/docs/APP_LSC_Sample_Preparation_Solubilization.pdf. Downloaded on May12, 2020.

Petersen HI. 1966. Translocation of ^{14}C-labelled 2,4-D in cereals. In *Isotopes in Weed Research Proceedings of the Symposium on the Use of Isotopes in Weed Research*, ed. International Atomic Energy Agency, 27–29. IAEA, Vienna, Austria.

Powles S, Preston C. 2006. Evolved glyphosate resistance in plants: Biochemical and genetic basis of resistance. *Weed Technology*. 20: 282–289.

Rademachher B, Borner H, Morgenstern W, Rentschhler W. 1961. Unterscheidung von Unkrautkeimpflanzen durch vorherige Markierung der Samen mit Schwefel-35, *Weed Research*. 1: 196–202.

Ricci RA, Chilosi G, Varcaccio G, Vingiani GB, Van Lieshout R. 1960. The gamma ray spectra of [65]Ni and [65]Zn; characteristics of the lower excited states of [65]Cu. *Nuovo Cimento*. 17, 523–534.

Riesen O, Feller U. 2005. Redistribution of nickel, cobalt, manganese, zinc, and cadmium via the phloem in young and maturing wheat. *J Plant Nutr*. 28: 421–430.

Röntgen W. 1895. Sitzungsberichte würzburger physik-medic. *Gesellschaft* 137:132–141.

Rutherford E. 1911. The scattering of α and β particles by matter and the structure of the atom. *Philosophical Magazine* 21(125):669–688.

Slade P. 1966. Possible appearance of degradation products of paraquat in crops. In *Isotopes in Weed Research Proceedings of the Symposium on the Use of Isotopes in Weed Research*, ed. International Atomic Energy Agency, 113–120. IAEA, Vienna, Austria.

Süss A, Schurmann G, Pfann G. 1966. Behaviour of ^{14}C-labelled diesel oil in different soils. In *Isotopes in Weed Research Proceedings of the Symposium on the Use of Isotopes in Weed Research*, ed. International Atomic Energy Agency, 125–133. IAEA, Vienna, Austria.

Temple S. 2015. Liquid scintillation counting: how has it advanced over the years and what does the future hold? *Bioanalysis* 7(5):503–505.

Venescu RE, Valeca M, Bujoreau L, Bujoreau D, Venescu B. 2016. C-14 analysis in radioactive waste by combustion and digestion. *Nuclear*. 2016: 118–125.

Wakelin AM, Lorraine-Colwill DF, Preston C. 2004. Glyphosate resistance in four different populations of *Lolium rigidum* is associated with reduced translocation of glyphosate to meristematic zones. *Weed Research*. 44: 453–459. doi:10.1111/j.1365-3180.2004.00421.x

Wehtje G., Miller M.E., Grey T.L. and Brawner W.R. 2007. Comparisons between X-ray film- and phosphorescence imaging-based autoradiography for the visualization of herbicide translocation. *Weed Technology* 21(4):1109–1114.

Woods Hole Oceanographic Institution. 2007. NOSAMS: Radiocarbon Data Calculations. https://www.whoi.edu/nosams/radiocarbon-data-calculations. Retrieved 21 May 2020.

Yamaguchi S, Craft AS. 1958. Autoradiographic method for studying absorption and translocation of herbicides using ^{14}C labelled compounds. *Hilgardia*. 28: 161–91.

Yanniccari M, Istilart C, Giménez DO, Castro AM. 2012. Effects of glyphosate on the movement of assimilates of two *Lolium perenne* L. populations with differential herbicide sensitivity. *Environmental and Experimental Botany*. 82: 14–19.

Zuo X, Lu H, Jiang L, et al. 2017. Rice domestication began about 10,000 years ago. *Proc Natl Acad Sci*. 114: 6486–6491. doi: 10.1073/pnas.1704304114

Zweep WVD. 1961. The movement of labelled 2,4-D in young barley plants. *Weed Research*. 1: 258–266.

chapter two

Sorption and desorption studies of herbicides in the soil by batch equilibrium and stirred flow methods

Kassio Ferreira Mendes[1], Rodrigo Nogueira de Sousa[2], Matheus Bortolanza Soares[2], Douglas Gomes Viana[2], Adijailton Jose de Souza[2]
[1]*Federal University of Viçosa*
[2]*University of São Paulo*

Contents

2.0   Introduction

The environmental behavior of herbicides is closely related to their availability in the soil to control weeds. The plant–soil–herbicide interaction alters the availability of the herbicide in the soil solution and is governed by the processes of retention and absorption of these compounds by weeds (Takeshita et al. 2019a). When the herbicide is applied pre-emergence, directly to the soil, it is necessary that the product is bioavailable in the soil solution, not absorbed or in the form of bound residue, so that it is absorbed by weeds and has efficient control (Chi et al. 2017). Sorption process also influences the herbicide dissipation processes in the environment, such as degradation, volatilization, leaching, and runoff (Herwig et al. 2001, Gavrilescu 2005).

Sorption refers to the capacity of the soil to retain the herbicide molecule, decreasing its availability in the soil solution. However, herbicides have two properties: (i) neutral and (ii) ionizable (weak acids and bases) that can affect sorption. In nonionic herbicides, sorption can occur by hydrogen bonds, van der Waals forces, and hydrophobic partition (Berkowitz et al. 2008). Hydrogen bridges are formed by bonds between two strongly electronegative atoms and are often found in herbicides that have functional groups with C = O bonds (e.g. isoproturon) linked to SOM, and clay minerals from oxygen-containing radicals (-O-2) and hydroxyls (-OH) (Spark and Swift 2002). Van der Waals forces occur through induced dipole-dipole or dipole interactions, generating a weak, short-range attraction between the solute and the sorbent. Therefore, this mechanism is reported in the sorption of several non-ionic molecules, including alachlor (Lenheer and Aldrichs 1971, Senesi et al. 1986). The hydrophobic partition occurs regardless of the pH of the medium for herbicides that have low solubility of hydrophobic character. These molecules tend to interact weakly with hydrophobic sites in SOM, becoming the main sorption pathway for several herbicides (metolachlor, diuron, among others) (Wauchope et al. 2002). Even so, the sorbed molecules can return to the soil solution through the desorption process, depending on the physical-chemical properties of each herbicide (Koskinen and Harper 1990). The sorption and desorption processes are complex between the soil solution (liquid phase) and the soil colloids (solid phase) and are outlined in Figure 2.1.

In some situations, the sorbed herbicide molecules may pass into forms unavailable in the soil, called bound waste. The formation of bound residue is an important mechanism for the dissipation of herbicides and can remain in the soil for a long time (Gevao et al. 2000). However, in specific cases, part of this fraction linked to the soil may become available to weeds, a process called remobilization (Amondham et al. 2006), according to Figure 2.1.

In general, the physical and chemical properties of herbicides that influence their soil retention process are:

- Octanol-water partition coefficient (K_{ow})
- Coefficients of partition on the ground or sorption–desorption (K_d)
- Acid/base dissociation coefficients (pK_a or pK_b)
- Degradation in the environment (half-life time, DT_{50})
- Coefficient of air–water partition or constant of Henry's law (K_H)

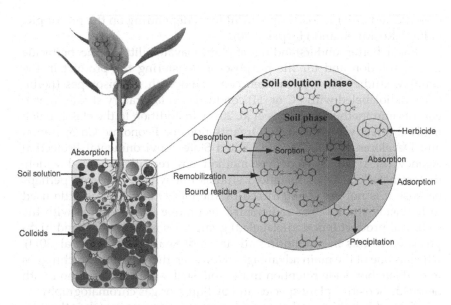

Figure 2.1 Process of sorption and desorption of herbicides in the soil matrix and absorption by weeds.

- Vapor pressure (PV)
- Water solubility (S$_w$)

The herbicide sorption and desorption process is influenced by several physicochemical properties of the soil, such as the content of soil organic matter (SOM), cation exchange capacity (CEC), iron and aluminum oxides, texture and pH, in addition to environmental conditions such as rainfall, temperature and humidity (Liu et al. 2010). SOM plays the most important role because it serves as a source of energy and nutrients to microorganisms responsible for herbicide degradation, and has active sites for the retention process, such as structural stabilizer and chemical buffer for herbicides, immobilizing them. Most herbicide sorption studies have been related to the organic carbon (OC) contents that influence sorption in a similar way to the content and quality of the clay fraction, and is also the main site of formation of bonded residue of the products (Mendes et al. 2014). Texture is another factor that affects herbicide sorption, as previously reported, causing soils with high clay levels to have higher sorptive capacity, so the dose of herbicides recommended in pre-emergence is higher in clay soils when compared to sandy soils. At the same time, the pH value of the soil solution also influences the sorption of ionizable herbicides (acids or bases), since these products in the soil solution can vary the amount of cationic and anionic loads or neutral form depending on the pK_a or pK_b values (Koskinen and Harper 1990).

For a better understanding of the bioavailability of the herbicide in soil solution and for the purpose of registering the product in the country, studies have often been conducted with radioisotopes (herbicides radiolabeled with ^{14}C or ^{3}H) conducted in laboratory studies (batch equilibrium method) (Mendes et al. 2017). In addition to the classic batch method already regulated by Organisation for Economic Co-operation and Development (OECD) and United States Environmental Protection Agency (EPA), the stirred flow method has recently emerged widely disseminated in metal/metalloid studies but which, however, perhaps because it is not yet standardized by the OECD and EPA, is little used in herbicide studies. The radiometric technique in association with the technical product (non-radiolabeled) guarantees relatively fast and high-precision results in monitoring all stages of analysis (Wang et al. 2014), which is one of the main advantages. However, there are other techniques to evaluate herbicide retention in the soil, such as bioassay method with herbicide-sensitive plant species and by liquid or gas chromatography.

To establish the sorption and desorption isotherms, which is the relation of herbicide concentrations in the solution and in the solid phase of the soil, different increasing concentrations of herbicides, usually five, are used, depending on the recommended dose for the crop in the field (Limousin et al. 2007). The equation of linear isotherm allows the calculation of the partition coefficient (K_d) and partition in relation to the OC content (K_{oc}) frequently used by the researchers (Braun et al. 2017). Knowledge of sorption isotherms is important in soil and weed science for predicting herbicide activity for weed control, soil transport and persistence, and carryover for crops in rotation and/or succession (Pereira et al. 2018).

Although the OECD (2000) and EPA (2008) describe standards for the sorption and desorption studies of radiolabeled herbicides with ^{14}C and ^{3}H, researchers in this area still lack information, mainly with herbicides. Therefore, in this chapter subsidies regarding the factors intrinsic to herbicide molecules and soil characteristics that control the process of sorption and desorption of herbicides will be provided in addition to the main methods for the study of the kinetics of chemical reactions in herbicides radiolabeled, as well as the theoretical basis of kinetics and construction of sorption and desorption of these radioisotopes in soils.

2.1 Concepts

Retention: Soil capacity to sorb and desorb the herbicide, decreasing its availability in soil solution and being able to return to its herbicide activity (Koskinen and Harper 1990). Thus, the retention process involves the sorption and desorption of the herbicide to the soil colloids.

Sorption: Term used to describe the retention process in general, without distinction to the specific processes of absorption, adsorption, and/or precipitation. Sorption is an interfacial process and refers to the adhesion or attraction of one or more ionic or molecular layers to a surface (Sposito 2016).

Absorption: A term used when herbicide molecules penetrate the pores of soil aggregates or when these products are immobilized by the microbiota and plants (Chiou 2002).

Adsorption: An accumulation process in an interface that is contrasted with absorption or passing through an interface. Adsorption is one of the most important mechanisms that reduces herbicide concentration in soil solution (Chiou 2002).

Precipitation: Occurs when the concentration of molecules in the soil solution is equal to or greater than the solubility of the herbicide. Thus, precipitation refers to the passage of molecules from the liquid phase to the solid phase.

Desorption: A process in which the herbicide molecule is released in whole or in part to the soil solution, promoting a new balance in the soil solution, being bioavailable (Barrow 1979).

Bound residue: When the herbicide desorption does not occur and the molecule remains sorbed after successive solvent extractions; that is, it is not bioavailable and may remain inactive for a long time (Boesten 2016).

Remobilization: When part of the bound residue returns to the soil solution over time, a process similar to desorption (Schäffer et al. 2018).

Mass balance: The percentage of herbicide recovered analytically after a retention study, divided into the product found as sorbed, desorbed, and bound residue in relation to the nominal amount of herbicide initially applied (Kniss et al. 2011).

Amount of sorbed and desorbed: The percentage of herbicide that was sorbed and desorbed, respectively, in relation to the amount initially applied (Mamy and Barriuso 2007).

Equilibrium time: The time when the herbicide concentration in the soil solution is in balance with the amount of sorption in soil particles (Koskinen and Harper 1990).

Equilibrium concentration (Ce): Herbicide concentration in soil equilibrium solution or humemic acid (EPA 1999).

Soil concentration (Cs): The difference between the initial herbicide concentration in the soil solution and the concentration present after equilibrium time.

Sorption and desorption isotherms: Mathematical equations used to conveniently describe the processes of sorption and desorption of herbicides (solutes) by the soil matrix in quantitative terms. The isotherms are obtained by the graphic relationship between the sorption/desorption concentration of herbicide in the solid phase and that remaining in the solution after reaching the balance with the soil (Sparks 2003).

Partition coefficient or sorption and desorption (K_d): Represents the relationship between the concentration of the herbicide sorbed (Cs) in the soil and its concentration in the equilibrium solution (Ce) (Schwarzenbach et al. 1993). The higher the K_d of a herbicide, the greater its soil retention capacity. There are several mathematical models to estimate these coefficients as a function of the concentrations applied in the soil.

Coefficient of sorption and desorption normalized by OC content (K_{oc}): The partition coefficient of the herbicide between soil-solution corrected by soil OC content (Schwarzenbach et al. 1993). This is commonly done for the coefficients of the other herbicide sorption and desorption models.

Hysteresis (H): A phenomenon in which part of the sorbed herbicide returns to the soil solution, and the desorption velocity of the soil colloid molecule is different (usually lower) than the sorption velocity (Celis and Koskinen 1999). Thus, there are negative, neutral, and positive hysteresis.

Sorption and desorption kinetics: The study of the reaction velocities that control the sorption and desorption processes of herbicides in the soil (Sparks 2003).

N-octanol-water partition coefficient (K_{ow}): The ratio of the concentration of an herbicide in the saturated n-octanol phase in water and its concentration in the saturated aqueous phase in n-octanol. K_{ow} values have no unit and are usually expressed in log K_{ow} (Mackay et al. 1997, Amézqueta et al. 2020). This coefficient measures the bioaccumulation of herbicides in the food chain.

Solubility in water (S_w): The abundance of the herbicide in the aqueous phase, when the solution is in balance with the pure compound in its state of aggregation the specific temperature (25°C) and pressure (1 atm) (Schwarzenbach et al. 1993).

Acid/base dissociation constant (pK_a/pK_b): The dissociation potential of an acid or basic herbicide, respectively, in liquid medium, in relation to the pH of the medium (Tan 2011). Thus, pK_a/pK_b is influenced by the pH value, in which when the $pK_a/pK_b = pH$, half of the molecules are dissociated and half are in molecular form (Figure 2.2). Basic herbicides present more than 90% of their molecules in molecular form when the pH value of the soil is greater than pK_b. On the other hand, acid herbicides present more than 90% of their molecules in molecular form when the pH value of the soil is lower than pK_a.

Half-life of degradation (DT_{50}): The ability of the herbicide to undergo a chemical, luminous, or biological reaction and turn into metabolites

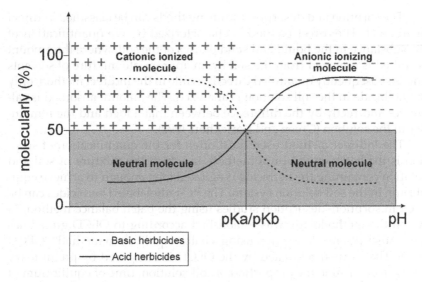

Figure 2.2 Scheme of dissociation of an acid or basic herbicide in the soil matrix.

and/or be mineralized. Thus, the DT_{50} of an herbicide can be defined as the time required for the initial concentration of the product to reduce to 50% (Schwarzenbach et al. 1993).

Vapor pressure (PV): The pressure exerted by a steam in equilibrium with a liquid, at a certain temperature. This characteristic indicates the degree of volatilization of the molecule, that is, its tendency to be lost to the atmosphere in the form of gas. The higher the pressure, the greater the potential for herbicide volatilization. Herbicide volatility increases under conditions of high temperature and low relative humidity (Schwarzenbach et al. 1993).

2.2 *Preliminary study in sorption and desorption studies*

The sorption–desorption process is dynamic in time and space, and in a way, it is in an equilibrium state in which the molecules are being continuously transferred from the liquid solution to the surface of the solid phase (Helling and Turner 1968). The study of this process makes it possible to know the nature of the herbicide–soil bond in addition to quantifying the herbicide retention in the soil. The analyses that propose to investigate these connections make it possible not only to know the isolated effects of each soil property on the retention of a certain herbicide in the soil, but the possible interactions of the various factors in a single model, since the soil properties are closely related (Oliveira Jr. et al. 2013, Takeshita et al. 2019b).

The sorption and desorption study methods can be classified as direct or indirect. The direct method is characterized by the quantification of the substance in the liquid and solid phases of the soil, where the amount absorbed is removed from the soil by an appropriate mixture of solvents and is subsequently quantified directly, while in the indirect method, only the herbicide in the soil solution is analyzed. The amount absorbed is calculated indirectly by the difference between that added and the remainder in the solution (Green and Karickhoff 1990).

The indirect method used most often for the quantification of sorption is the batch equilibrium method, in which the mixture of soil and solution containing the herbicide is agitated long enough to achieve equilibrium in the soil-solution system. The ^{14}C-radiolabeled herbicides can be used in sorption–desorption studies using the batch balance method, in which the methodology was established according to OECD guidelines 106, "Adsorption – desorption using a batch equilibrium method" (OECD 2000). The method regulated by the OECD is composed by preliminary study to determine the proportion of soil-solution, time of equilibrium of the evaluated substance (herbicide), and the sorption and stability of the herbicide during the study period. The studies of sorption and desorption of herbicides comprises three tiers (preliminary study, screening test, and determination of sorption and desorption isotherms) as represented in Figure 2.3.

Characterization, selection, collection, and storage of soil samples for sorption and desorption studies of herbicides are detailed by the guidelines of OECD (2000).

First, to conduct the preliminary study, it is necessary to add at least 45 mL of 0.01 mol L^{-1} CaCl$_2$ to the dry soil samples 12 h before the start of the study. The amount of stock solution should not exceed 10% of the final volume (50 mL). To represent the soil solution in real field conditions, 0.01 mol L^{-1} CaCl$_2$ is used and also serves to make the solution translucent at the time of collection (Mendes et al. 2017).

In the preliminary study, it is also necessary to use at least two soils with different textures (sandy and clayey) and with three different soil-solution proportions. The following are some soil-solution proportions (OECD 2000):

- 1:2, with 25 g of soil and 50 mL of herbicide solution
- 1:10, with 5 g of soil and 50 mL of herbicide solution
- 1:25, with 2 g of soil and 50 mL of herbicide solution

The initial solution must be prepared with the herbicide in the highest dose to be used in the study of sorption with 0.01 mol L^{-1} CaCl$_2$. Such a solution should be placed in the Teflon flasks and mechanically stirred on a horizontal shaking table until the sorption balance is achieved. The time

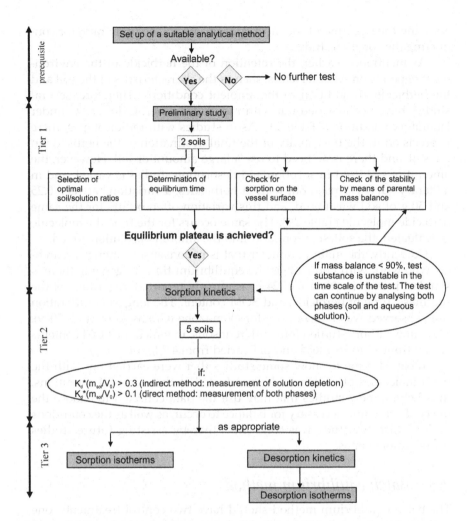

Figure 2.3 Sorption and desorption studies of herbicides comprises three tiers (preliminary study, screening test, and determination of sorption and desorption isotherms). (From OECD 2000.)

for equilibrium to be achieved is highly variable, depending on which herbicide is being studied in which soil type; however, the 24 h time is usually sufficient to achieve equilibrium. Equilibration time is the time the system needs to reach a plateau. At the end of the equilibration time, the vials should be centrifuged. In the case of the preliminary study, aliquots of 50 to 100 µL should be collected sequentially from the supernatant over a period of up to 48 h. And at least 20% of the herbicide applied must be absorbed into the soil, preferably between 50% and 80%, thus allowing the selection of the most appropriate soil-solution ratio for conducting the sorption study.

As mentioned earlier, the retention of the herbicide in the environment depends on several factors, such as the characteristics of the soil and the herbicide, in addition to the ambient conditions. Therefore, several studies have been carried out with radiolabeled herbicides at ^{14}C under laboratory conditions (Table 2.1). As in studies with radioisotopes, there is precision in the traceability of the final destination of the herbicide in the soil, and there is no need to use a large amount of soil. However, the amount of soil used in sorption and desorption studies is varied, and in a literary survey carried out with 21 herbicides, a variation between 0.25 and 50 g of soil was found, and this variation often occurs for the same herbicide molecule (Table 2.1). The same occurs for the time the molecule is evaluated, the soil/solution ratio, and the amount of solution added.

Currently, the main parameter that is in consensus among researchers who work with herbicide is the equilibrium time; in general, there is standardization in the use of 24 h, as already reported, regardless of the soil mass, type of herbicide, and SOM content. The longest equilibration time observed was 7 days used for clomazone (Gunasekara et al. 2009). The same authors studied four different soils, in which the SOM contents varied from 12 to 26 g kg^{-1} and pH varied from 4.7 to 6.4.

Table 2.1 aims to show some studies that were carried out with the main molecules of ^{14}C-herbicides conducted in different soil conditions, thus helping new studies that may use the data collected to choose the molecule, the time necessary for balance to occur, as well as the conditions under which the study was conducted, thereby assisting future studies using radioisotopes.

2.3   Batch equilibrium method

The batch equilibrium method should have two control treatments: one with only the test substance (radiolabeled herbicide) at the maximum recommended dose of the herbicide by the manufacturer in the crops without the addition of soil, in the solution of CaCl$_2$ to 0.01 mol L^{-1}, which will serve to check the stability of the herbicide in the CaCl$_2$ solution and its possible sorption to the walls of the vials (commonly of Teflon type); the other control should have only the soil in the solution of CaCl$_2$ at 0.01 mol L^{-1}, without the herbicide, which will serve as background (BG) control; that is, it can detect soils contaminated with herbicides that may affect the results (OECD 2000). Both controls should be performed in duplicate (at least), in order to generate a minimum amount of radioactive waste residue at the end of the studies.

In the sorption study, aliquots of the radiolabeled solution should be transferred in duplicate to scintillation vial containing 10 mL of

Table 2.1 Preliminary data collection for study of sorption and desorption of ^{14}C herbicides in the soil

Herbicide	Radioisotope	Specific activity	Soil[a]	pH[b]	O.C.[c]	Mass[d]	Solution[e]	Soil: solution	Equilibrium time[f]	Evaluated times[g]	Reference
Alachlor	[^{14}C-URL]	999 kBq mmol^{-1}	Haplorthox	4.8	14.5	10	10	1:1	24	-	Oliveira Jr. et al. (2001)
			Quartzipsamment	6.3	27.8						
			Hapludult	4.3	3.5						
				4.5	5.8						
				4.6	74.5						
				4.6	17.4						
Aminocy-clopyrachlor	^{14}C-aminocy-clopyrachlor	~109 Bq mL^{-1}	Oxisol - Rhodic Hapludox	6.0	16.0	10	-	1:1	24	0, 2, 4, 8, 24 and 48	Oliveira Jr. et al. (2011)
			Ultisol	5.7	18.3						
			Oxisol - Typic Hapludox	5.9	20.5						
			Plintosol	6.2	21.7						
			Typic quartzipsaments	5.1	6.8						
			Arenic albaqualfs	5.4	5.0						
				4.9	10.2						
				6.1	17.9						
				5.2	16.3						
				6.5	6.4						
				5.7	9.0						
				7.2	6.5						
				6.5	6.1						
				6.0	10.6						
Aminocy-clopyrachlor	[pyrimidine-2-^{14}C]-aminocy-clopyrachlor	46 Bq mL^{-1}	Ferralsol	6.7	25.0	5	10	1:2	24	24	Mendes et al. (2020b)
			Hapludalf	6.0	28.0						
			Chernozems	6.2	20.0						
Atrazine	^{14}C-atrazine	160 MBq mmol^{-1}	Ultisol	6.5	28.0	2	20	1:1	24	24	Correia et al. (2007)

(Continued)

Table 2.1 Preliminary data collection for study of sorption and desorption of ^{14}C herbicides in the soil (Continued)

Herbicide	Radioisotope	Specific activity	Soil[a]	pH[b]	O.C.[c]	Mass[d]	Solution[e]	Soil: solution	Equilibrium time[f]	Evaluated times[g]	Reference
Atrazine	[14C-URL]	129 kBq mmol[-1]	Haplorthox Quartzipsamment Hapludult	4.8 6.3 4.3 4.5 4.6 4.6	1.4 2.7 0.3 0.5 7.4 1.7	10	10	1:1	24	-	Oliveira Jr. et al. (2001)
Clomazone	[ring-U-14C]-clomazone	3.83 µCi µmol[-1]	-	6.1 5.7 4.7 6.4	26.0 25.0 12.0 12.0	1-2	7	~1:4	7 days	7 days	Gunasekara et al. (2009)
Dicamba	^{14}C-dicamba	106 MBq mmol[-1]	Hapludox	6.7	15	10	10	1:1	24	16-20	Koskinen et al. (2006)
Dicamba	[14C-URL]	106 kBq mmol[-1]	Haplorthox Quartzipsamment Hapludult	4.8 6.3 4.3 4.5 4.6 4.6	1.45 2.78 0.35 0.58 7.45 1.74	10	10	1:1	24	-	Oliveira Jr. et al. (2001)
Diketonitrile	Phenyl-U-[14]C	901 MBq mmol[-1]	Hapludox	6.7	15	10	10	1:1	24	16-20	Koskinen et al. (2006)

(Continued)

Table 2.1 Preliminary data collection for study of sorption and desorption of ¹⁴C herbicides in the soil (Continued)

Herbicide	Radioisotope	Specific activity	Soil[a]	pH[b]	O.C.[c]	Mass[d]	Solution[e]	Soil: solution	Equilibrium time[f]	Evaluated times[g]	Reference
Diuron	[ring-U-¹⁴C]-diuron	332 MBq mmol⁻¹	-	6.2	19.0	10	10	1:1	24	72	Rubio-Bellido et al. (2016)
				6.9	12.0						
				8.0	10.0						
				8.7	10.0						
				5.5	9.0						
				6.0	8.0						
				8.5	1.0						
Glyphosate	¹⁴C-Glyphosate	0.67 kBq mL⁻¹	Inceptisol	5.7	9.0	0.5	1	1:2	24	24	Sharma and Lai (2019)
			Entisol	8.7	25.0	0.5		1:2			
			Ferralsol	5.6	44.0	0.25		1:4			
			Vertisol	7.8	22.0	0.5		1:2			
Hexazinone	[triazine-6-¹⁴C]-hexazinone	3.14 MBq mg⁻¹	Hapludalf	6.9	5.0	10	10	1:1	24	48	Mendes et al. (2019)
Hexazinone	[carbonyl-¹⁴C]	66 kBq mmol⁻¹	Haplorthox	4.8	1.4	10	10	1:1	24	-	Oliveira Jr. et al. (2001)
			Quartzipsamment	6.3	2.7						
			Hapludult	4.3	0.3						
				4.5	0.5						
				4.6	7.4						
				4.6	1.7						
Imazaquin	¹⁴C-imazaquin	0.54 TBq kg⁻¹	Umbraquult	4.7	28.0	10	20	1:2	24	4, 8 and 16	Weber et al. (2003)
			Hapludult	6.7	5.0						

(Continued)

Table 2.1 Preliminary data collection for study of sorption and desorption of ^{14}C herbicides in the soil (Continued)

Herbicide	Radioisotope	Specific activity	Soil[a]	pH[b]	O.C.[c]	Mass[d]	Solution[e]	Soil: solution	Equilibrium time[f]	Evaluated times[g]	Reference
Imazaquin	-	-	Haplustox	6.7	7.0	5	10	1:1	24	1, 5, 9, 24 and 48	Oliveira et al. (2006)
			Rhodic Haplustox	5.1	15.0						
Imazaquin	^{14}C-imazaquin	0.8 MBq mg^{-1}	Oxisol	6.1	11.0	5	10	1:2	24	0.5, 1, 3, 6, 12, 24 and 48	Barizon et al. (2005)
Imazethapyr	[^{14}C-URL]	214 kBq mmol^{-1}	Haplorthox	4.8	1.4	10	10	1:1	24	-	Oliveira Jr. et al. (2001)
			Quartzipsamment	6.3	2.7						
			Hapludult	4.3	0.3						
				4.5	0.5						
				4.6	7.4						
				4.6	1.7						
Indaziflam	triazine-2,4,-^{14}C	~72 Bq mL^{-1}	Oxisol - Rhodic Hapludox	6.0	1.6	4		1:2	24	0, 2, 4, 8, 24 and 48	Alonso et al. (2011)
			Oxisol - Typic Hapludox Arenic albaqualfs	5.9	2.0						
			Typic quartzipsaments	6.2	2.1						
				5.4	0.5						
				6.0	1.0						
			Mollisol - Typic calciaquolls	6.5	0.6						
			Mollisol -Typic hapludolls	8.1	2.1						
				8.3	1.1						
				6.0	2.5						

(Continued)

Table 2.1 Preliminary data collection for study of sorption and desorption of ^{14}C herbicides in the soil (*Continued*)

Herbicide	Radioisotope	Specific activity	Soil[a]	pH[b]	O.C.[c]	Mass[d]	Solution[e]	Soil: solution	Equilibrium time[f]	Evaluated times[g]	Reference
Metolachlor	^{14}C-metolachlor	0.96 TBq kg⁻¹	Hapludult Umbraquult	6.7 4.7	5.0 28.0	10	20	1:2	24	-	Weber et al. (2003)
Metolachlor	[U-ring-^{14}C]-metolachlor	2.40 Bq mL⁻¹	Ferralsol Luvisols Chernozems	6.7 6.0 6.2	25.0 28.0 20.0	5	10	1:2	24	24	Mendes et al. (2020b)
Metribuzin	[ring-6-^{14}C]-metribuzin	5.43 MBq mg⁻¹	Hapludalf	6.9	5.0	10	10	1:1	24	48	Mendes et al. (2019a)
Metsulfuron-methyl	[phenyl-^{14}C-URL]	154 kBq mmol⁻¹	Haplorthox Quartzipsamment Hapludult	4.8 6.3 4.3 4.5 4.6 4.6	1.4 2.7 0.3 0.5 7.4 1.7	10	10	1:1	24	-	Oliveira Jr. et al. (2001)
Nicosulfuron	[pyridine-2-^{14}C]	158 kBq mmol⁻¹	Haplorthox Quartzipsamment Hapludult	4.8 6.3 4.3 4.5 4.6 4.6	1.4 2.7 0.3 0.5 7.4 1.7	10	10	1:1	24	-	Oliveira Jr. et al. (2001)
Quinclorac	[2,3,4-^{14}C]-quinclorac	1.5 MBq mg⁻¹	Hapludalf	6.9	5.0	10	10	1:1	24	48	Mendes et al. (2019a)

(*Continued*)

Table 2.1 Preliminary data collection for study of sorption and desorption of ^{14}C herbicides in the soil (Continued)

Herbicide	Radioisotope	Specific activity	Soil[a]	pH[b]	O.C[c]	Mass[d]	Solution[e]	Soil: solution	Equilibrium time[f]	Evaluated times[g]	Reference
Simazine	[^{14}C-URL]	100 kBq mmol^{-1}	Haplorthox Quartzipsamment Hapludult	4.8 6.3 4.3 4.5 4.6 4.6	1.4 2.7 0.3 0.5 7.4 1.7	10	10	1:1	24	-	Oliveira Jr. et al. (2001)
Sulfometuron-methyl	[phenyl-^{14}C-URL]	157 kBq mmol^{-1}	Haplorthox Quartzipsamment Hapludult	4.8 6.3 4.3 4.5 4.6 4.6	1.4 2.7 0.3 0.5 7.4 1.7	10	10	1:1	24	-	Oliveira Jr. et al. (2001)
Tebuthiuron	[^{14}C-UL]-tebuthiuron	3.01 MBq mg^{-1}	Oxisol	4.5 4.3	8.0 23.0	5	10	1:2	24	21 days	Toniêto et al. (2016)
2,4 D	[^{14}C-ring-labeled]-dichlorprop	55.01 µCi/mg	-	4.6	34.0	0.5	5	1:10	24	4 weeks	Riise and Salbu (1992)

a Soil classification based on WRB/FAO and Soil Taxonomy/USDA.
b Soil pH determined in water.
c Organic carbon content in the soil (g).
d Mass of soil used in the experiment (g).
e Volume of solution used in the experiment (mL).
f Equilibrium time. There is no information in the article.
g Time evaluated in the study (hours). There is no information in the article.

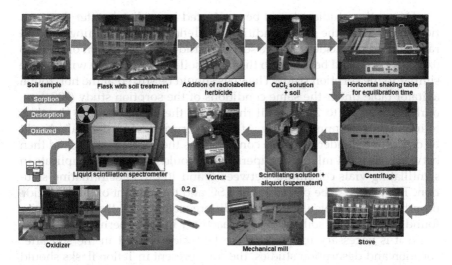

Figure 2.4 Representation of sorption and desorption studies of radiolabeled herbicides by batch method in equilibrium.

scintillating solution (cocktail), and these vials should be placed in the liquid scintillation counter (LSC) for at least 5 min in order to measure the initial amount of radiolabeled herbicide measured by the radioactivity of the solution obtained in the reading. The entire procedure for the sorption and desorption studies of herbicides radiolabeled by batch method is presented in the Figure 2.4.

Also in duplicate, 10 mL of the radiolabeled solution of all concentrations (at least five concentrations) should be added to Teflon vials containing soils with the soil-solution ratio already defined in the preliminary study, which is usually 1:10. The vials should be stirred on the horizontal agitator table in a dark room (to eliminate photodegradation) with temperatures of 20 to 25°C, depending on the objective of the study, at approximately 200 rpm for the solution to reach equilibrium in the time found in the preliminary study.

At the end of the equilibrium time, the vials should be centrifuged between 3000 and 3500 rpm for 15 min, then between 0.5 and 1 mL of the supernatant of each vial should be added in duplicate in flicker bottles containing between 5 and 10 mL of scintillating solution. The vials should be placed in the LSC, and the amount of radiolabeled herbicide will be determined by reading the radioactive activity. The sip of the herbicide will be calculated by indirect method by means of the difference between the initial radioactivity applied and the radioactivity obtained in the supernatant after equilibrium.

Desorption studies should be conducted immediately after the sorption study under the same conditions. In Teflon vials containing soil and radiolabeled herbicide from the sorption study, a new solution of $CaCl_2$ to 0.01 mol L^{-1} should be added to the moist soil. The procedure will now be similar to the sorption study. The vials should be stirred at the horizontal agitator table under the same conditions of the sorption study (200 rpm, dark room of 20 to 25°C) until they reach the equilibrium in the time found in the preliminary study. After the stirring period, the vials should be centrifuged under the same conditions as the sorption study and then between 0.5 and 1 mL of the supernatant should be added in duplicate in scintillating vials containing between 5 and 10 mL of scintillating solution. The vials will be placed in the LSC and the amount of the herbicide desorbed will be calculated by the difference between the radioactivity found sorbed to the soil and the radioactivity in the supernatant.

If it is necessary to quantify the herbicide present in the soil after sorption and desorption studies, the soils present in Teflon flasks should be dried in a stove at 40°C for about 48 h. After drying, they should be milled, and if necessary, can be stored in plastic bags. Subsequently, three replicates of 0.2 g of each soil should be oxidized in the oxidizer in order to determine the amount of radioactive herbicide still present in the soil. It is noteworthy that this stage of the procedure is not common in studies of sorption and desorption of herbicides in the soil.

The mass balance will be obtained by means of the proportion between the sum of the radiolabeled herbicide recovered in all stages (sorbed, desorbed, and soil oxidation) and the amount of the radiolabeled herbicide initially applied. The recovery percentage should be between 90% and 110%, so that the data are accepted by quality control in studies with radiolabeled molecules (OECD 2000).

2.4   Stirred flow method

The agitated flow reactor is International Union of Pure and Applied Chemistry (IUPAC) as a reactor in which an effective mixture is achieved, placing the catalyst in a fast-rotating chamber (Burwell 1976). The stirred flow method is not listed in the OECD and EPA as a standard method for analyzing herbicide sorption and desorption kinetics (EPA 2008, OECD 2000). However, due to some limitations present in the balance batch method, such as soil:solution ratio, where successive dilution steps result in re-sorption that can prevent desorption (Sparks, 2003), the use of the stirred flow method arose for evaluating the kinetics of chemical reactions in herbicides (still incipient) and metals (already widely used) (Tuin and Tels 1991, Peng et al. 2018).

The stirred flow method consists of using a chamber in which the mixture soil:solution is stirred continuously. A solution input stream is constantly maintained while the outflow is collected using a fraction collector. There are some similarities in the stirred method with the column method, however the main differences between the two methods are in the soil:solution and hydrodynamics ratio (Martinez et al. 2010). The low proportions of solids and liquids in the stirred flow method result in the arrival of effluent solute faster than in soil columns (Sun and Selim 2019). In hydrodynamics, the movement of the solution in the stirred flow is considerably simpler than the transport of the solution by advection-dispersion in soil columns (Heyse et al. 1997). Another advantage of the stirred flow method over discontinuous methods is that relatively fast reactions can be measured, and the sorption and desorption kinetics can be measured in a single experiment (Strawn and Sparks 2000).

The stirred flow method generally consists of a cylindrical polypropylene micro reactor (~1.5 cm^{-3}), with a lateral inlet on the bottom and an outlet on the lid, both covered with 0.45 polytetrafluoroethylene (PTFE) filters µm (Ø 10 mm) to retain the soil in the reactor. The inlet is connected to a peristaltic pump by a PTFE tubing of approximately 0.5 mm. The reactor outlet tubing is connected to an automatic fraction collector, where effluent fractions are collected in collector blocks of approximately 2 mL. Stirring can be provided by a magnetic bar rotating at 400 rpm (Figure 2.5).

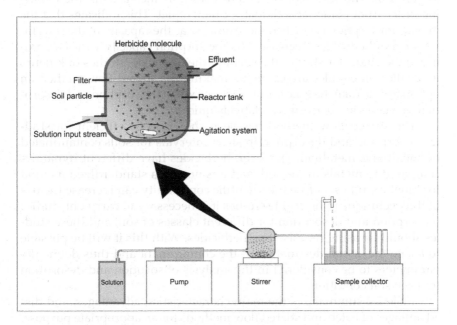

Figure 2.5 Schematic diagram of operation for stirred flow method.

As the stirred flow method is not a method standardized by public institutions (OECD and EPA), the analysis procedures can vary in the same way that they vary in batch studies, where the volumes of the aliquots vary from a few milliliters (~5 mL) when the herbicide is not radiolabeled (Tsai and Chen 2013) at a fraction of a milliliter (~200 μL) when radiolabeled isotopes are used (Mendes et al. 2017). The most recent study involving the use of the stirred flow method to assess the retention kinetics of unmarked herbicides was carried out by Conde-Cid et al. (2017), in which the authors evaluated the retention of quaternary ammonium herbicides in acidic soils.

The study was carried out using the herbicides paraquat, diquat, and difenzoquat, and the dose used for all herbicides was 0.1 mmol L^{-1} with the addition of 10 mmol L^{-1} of $CaCl_2$ as a supporting electrolyte (Conde-Cid et al. 2017). To perform the desorption of the herbicides, 10 mmol L^{-1} of $CaCl_2$ was used with a pH maintained close to 4.5. Outflow solutions were collected every 6 min in the sorption and desorption steps. The soils used were classified as Regosols (IUSS Working Group WRB 2014) and the main difference between the soils was the OC content. Soil 1 had 3.1 g kg^{-1} of OC; soil 2, 47.6 g kg^{-1} of OC; soil 3, 37.3 g kg^{-1} of OC; and soil 4, 12.5 g kg^{-1} of OC.

The main results indicate the greater sorption capacity of soil 1 for all herbicides, followed by soil 2, and lower values for soil 3 and soil 4 (Figure 2.6). The values of the kinetic constants increased in the following order from paraquat > diquat > difenzoquat. This indicates that the maximum sorption is reached more quickly as the capacity of the soils for sorption of herbicides decreases. As the sorption capacity varies according to the characteristics of the soil and the herbicide, studies of kinetics in smaller time scales are essential, making the stirred flow method an option to elucidate the kinetics of reactions when the sorption and desorption processes are performed relatively quickly.

The stirred flow method has been shown to be efficient for studies of sorption and desorption in short intervals for soils contaminated by metal and metalloid. However, herbicides have different dynamics compared to metals in the soil, and it is not yet a standardized method for kinetic studies. Before the scientific community can increase the use of this technique in herbicide studies it is necessary to carry out studies on sorption and desorption for different classes of soil, and these studies should be using radioisotope herbicides. With this it will be possible to locate the actual destination of the contaminant and thus define the parameters to be considered in the analysis of sorption and desorption for each soil condition.

Table 2.2 summarizes the main characteristics, advantages, and disadvantages of batch and stirred flow methods for an appropriate purpose. The batch method indicated a fast sorption, but it prevented concluding

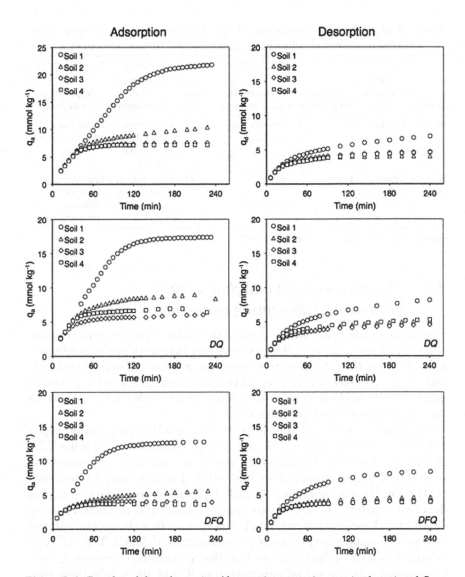

Figure 2.6 Results of the adsorption/desorption experiments in the stirred flow chamber for paraquat (PQ), diquat (DQ) and difenzoquat (DFQ). q_a: accumulated amount of herbicide adsorbed; q_d: accumulated amount of herbicide desorbed. Soil 1: low organic carbon (OC) content; Soil 2 and 3: high OC content; and Soil 4: medium OC content. (From Conde-Cid et al. 2017.)

to a slow reaction because of long-term side effects, such as particle disrupting. On the contrary, the stirred-flow through method used with two different mean water residence times confirmed the existence of slow reactions (Limousin et al. 2007).

Table 2.2 Main characteristics of experimental methods (batch and stirred flow) for the measurement of sorption–desorption isotherms and kinetics

Method	Solid/ solution ratio	Sorption measurement	Desorption measurement	Inert tracer	Main advantage/ Main disadvantage	Appropriate purpose
Batch	Chosen ($<1\ kg\ L^{-1}$)	Easy	Difficult (step by step)	No	Easy-to-use/ Unrealistic for porous media	Preliminary measurements
Stirred flow	Chosen ($<1\ kg\ L^{-1}$)	Easy	Easy	Possibly no	Flow-through with chosen soil/Solution ratio	Reaction kinetics, release measurement

Source: Adapted from Limousin et al. (2007).

2.5 Classification and modeling of the sorption and desorption isotherms

The interaction between herbicides and the surface of mineral and organic colloids of soils can be described using desorption and sorption isotherms. They describe the relationship between the mass of the sorbed herbicide and its concentration in the equilibrium solution. The term isotherm derives from constant temperature, which is one of the requirements for the evaluation of sorption by this method.

Desorption and sorption isotherms are subdivided into four types (Giles et al. 1974): type C, L, H, and S. Type isotherm "C" (Figure 2.7a) describes that the relationship between herbicide concentration in solution (C_e) and sorbed in the soil (C_s) is the same in any concentration. This relationship is called partition coefficient or K_d, as described earlier. The use of this type of isotherm is restricted to very small initial concentrations, when it is certain that the solute will not occupy all surface sorption sites (Limousin et al. 2007). The "L" type isotherm (Figure 2.7b) indicates that the relationship between C_e and C_s decreases as the initial concentration (C_i) of the solute increases. In other words, sorption sites are saturated so that the surface has a limited sorption capacity. Type isotherm "H" (Figure 2.7c) is a particular case of type "L" isotherm that describes a high initial affinity of the solute on the surface. The "S" type isotherm (Figure 2.7d) is a sigmoidal curve, i.e. it has an inflection point that is the result of at least two opposing sorption mechanisms. For example, the polar herbicides have low affinity with the surface of clays. However, as the clay surface is covered by these molecules, others are retained more easily (Pignatello 2000).

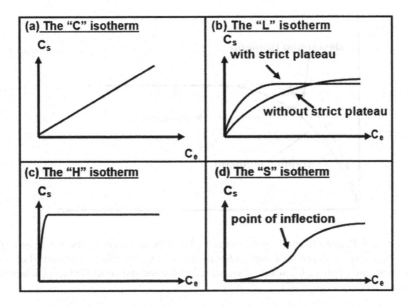

Figure 2.7 Graphical description of isotherms (a) "C", (b) "L", (c) "H", (d), and "S" type of sorption and desorption herbicide in the soil. C_e, herbicide concentration in equilibrium solution; C_s, solute concentration sorbed the solid surface. (Adapted from Giles et al. 1974 and Limousin et al. 2007.)

Weber and Miller (1989) analyzed data from 230 sorption isotherms of organic molecules in the soil and found the following distribution according to the classification of Giles et al. (1974): S = 16%; L = 64%; H = 12%, and C = 8%. This result indicates that the main factor that controls the shape of an isotherm is the characteristic of the adsorbent material (colloid surface). For example, fluridone (herbicide for aquatic weed control) sorption exhibited "S" isotherm in soils with low SOM content and high montmorillonite content, "L" isotherm in soils with high SOM content and low montmorillonite content, and isotherm type "C" in soils with low SOM and clay content (Weber et al. 1986).

Herbicide sorption kinetics is characterized by a fast and reversible stage followed by a slow and non-reversible stage or biphasic kinetics (Karickhoff and Morris 1985). The first phase comprises the sorption of the molecule in a labile form in the soil matrix and therefore can be easily desorbed or suffer microbial attack. In the second phase, the lower sorption–desorption velocity can be interpreted by diffusion of herbicide molecules in micropores of soil organomineral aggregates (Wu and Gschwend 1986, Steinberg et al. 1987, Ball and Roberts 1991).

It is assumed that a sorption isotherm is constructed based on data collected when the reaction reaches the chemical equilibrium. In this case,

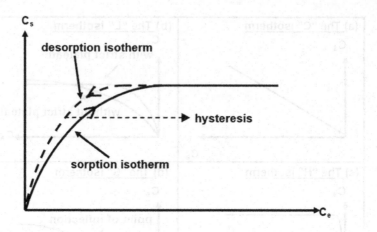

Figure 2.8 Representation of hysteresis phenomenon during the processes of sorption and desorption of herbicides in the soil. C_e, herbicide concentration in equilibrium solution; C_s, solute concentration sorbed the solid surface. (Adapted from Limousin et al. 2007.)

the sorption isotherm should have the same conformation as desorption isotherm, because the theory of thermodynamic equilibrium assumes the complete reversibility of the reactions at the same speed of product formation (Limousin et al. 2007). However, herbicide sorption kinetics involves the hysteresis (H) phenomenon; i.e., the desorption velocity of the soil colloid molecule is different (usually lower) than the sorption speed (Figure 2.8).

Hysteresis occurs due to the heterogeneity of the surface sorption sites and consequently the various sorption mechanisms between the herbicide and the soil. Even in the first phase of sorption, many sites exhibit the characteristic of irreversibility; that is, the molecule is strongly retained which prevents the desorption process from occurring (Celis and Koskinen 1999).

This non-singularity or hysteresis in the desorption process can be evaluated by the hysteresis index (H = $1/n_{desorption}/1/n_{sorption}$), which establishes a relationship between the degree of linearity of sorption and desorption. When sorption is completely reversible the value of the H index will be close to 1 (0.7 > H > 1.0); that is, hysteresis does not occur (Barriuso et al. 1994). The lower this index, the greater the H phenomenon and consequently the herbicide sorption will have greater irreversibility (Barriuso et al. 1994, Seybold and Mersie 1996).

The main causes related to H index are devices related to the method, chemical or biological transformations of the original compound, irreversible or resistant fractions of sorbed compounds, and impossibility of establishing chemical balance between the sorbed phase and solution

(Pignatello 2000). For hydrophobic organic compounds, the main mechanism involved in hysteresis is the diffusion of the solute through the SOM structure, with the rate of desorption being slower than the sorption kinetics (Huang et al. 2003). Among the possible diffusion mechanisms, the one related to SOM has proved extremely important in the sorption of organic compounds, such as non-ionizable herbicides.

2.6   Models of sorption–desorption isotherms

The sorption isotherm models allow us to evaluate the effect of herbicide concentration on the herbicide sorptive behavior in the studied soil. For the use of these models in studies of sorption–desorption of radiolabeled herbicides, at least five concentrations of the herbicide studied should normally be used, as previously reported.

The sorption–desorption coefficients are determined by the Freundlich isotherm model:

$$C_s = K_f \times C_e^{1/n}$$

Where:

- C_s is the amount of herbicide sorbed into the soil (μmol kg^{-1})
- K_f is the Freundlich sorption–desorption coefficients (μmol$^{(1-1/n)}$ L$^{1/n}$ kg^{-1})
- C_e is the concentration of herbicide in the equilibrium solution (μmol L^{-1})
- $1/n$ is the degree of linearity of the isotherm

The expression of Langmuir isotherm is represented by the equation:

$$C_s = (q_{max} \times K_l \times C_e)/(1 + q_{max} \times C_e)$$

Where:

- C_s is a concentration of herbicide sorbed in the soil (mg kg^{-1})
- q_{max} is the maximum herbicide concentration when the surfaces of the sites are saturated (mg g^{-1})
- K_l is the sorption–desorption coefficients of Langmuir (L mg^{-1})
- C_e is the concentration of the herbicide in equilibrium in the soil solution (mg L^{-1})

The linear coefficient of sorption–desorption is also calculated (K_d, L kg^{-1}).

$$K_d = C_s/C_e$$

Sorption–desorption values are also presented in terms of an "apparent distribution coefficient" ($K_{d, app}$) calculated from mass balance considerations in relation to K_d obtained in equilibrium batch studies for a single concentration (Rahman et al. 2004), and the sorption–desorption coefficients normalized by OC content of the soil (K_{oc}, L kg^{-1}):

$$K_{oc} = (K_d/(\%OC)) \times 100$$

K_{foc} and K_{loc} values are also calculated similarly to K_{oc}. Herbicide desorption coefficients are determined similarly to sorption coefficients, in which the remaining amount of herbicides sorbed in the soil returns to the equilibrium concentration, being desorbed.

There are several other models in the scientific literature that can be described to calculate the sorption and desorption isotherms of herbicides in the soil, but the three models described above are the most used in studies of sorption–desorption of herbicides in the soil. To check the other models, you can consult Ayawei et al. (2017) and Limousin et al. (2007), because the authors conducted a general review of the applications of various sorption–desorption isotherms in scientific studies.

2.7 Examples of sorption and desorption studies of herbicides in soil

Sorption–desorption studies are crucial to understand the interaction of the herbicide molecule with a soil matrix, as well as to infer the potential risk of contamination of water resources, biota, humans, in addition to allowing the safe and efficient use of the herbicide (Oliveira and Brighenti 2011, Mendes et al. 2017).

The use of the ^{14}C radioisotope in batch method studies has been the main approach to predict the several behavior of herbicides in the environment (Mendes et al. 2017), due to the simplicity of conducting the tests. This type of study provides information about the sorption and desorption coefficient, isotherms, and how the target molecule interacts with abiotic and biotic attributes of the environment (Oliveira Jr. and Regitano 2009).

There are extensive data available in the scientific literature regarding the environmental behavior of the main classes of herbicides. However, sorption–desorption studies are constantly required for monitoring and regulation purposes in the country of the use of this type of chemical (Gill and Garg 2014, Rigotto et al. 2010).

Several factors have been considered in classic sorption–desorption studies, such as properties inherent to the chemical group of herbicide, soil mineralogy, organic matter content, mixture of herbicides in tanks, soil conditioners and aging of the herbicidal molecule in the soil and more application technologies.

With the radiolabeled herbicide approach, Singh et al. (2014) evaluated how the variability of soil properties affected the sorption of ^{14}C-glyphosate, ^{14}C-2,4-D, and ^{14}C-atrazine in 140 soils from two humid landscapes in western Canada. Sorption coefficients (K_d) were widely variable in the three mineral horizons (A, B, and C). In general, ^{14}C-glyphosate showed higher K_d values, which were higher in horizon B of both collection sites, followed by K_d values of ^{14}C-atrazine and ^{14}C-2,4-D. The soil attributes that best correlated with an increase in the K_d value were soil organic carbon (SOC) and clay content. However, when evaluated in depth, the decrease in SOC content was negatively correlated with K_d for the three herbicides evaluated.

Bonfleur et al. (2016) evaluated the sorption of the herbicides ^{14}C-alachlor, ^{14}C-bentazone, and ^{14}C-imazethapyr in 11 Oxisols, with 6 soils from the subtropical region and soils from the tropical region under no-tillage. The authors observed higher K_d values in soils in the subtropical region [K_d (alachlor) = 1.04 to 3.04 L kg^{-1} and K_{oc} (alachlor) = 44.10 to 63.10 L kg^{-1}; K_d (bentazone) = 0.36 to 1.16 L kg^{-1} and K_{oc} (bentazone) = 9.50 to 24.10 L kg^{-1}; K_d (imazethapyr) = 0.99 to 4.64 L kg^{-1} and K_{oc} (imazethapyr) = 28.70 to 124.10 L kg^{-1}].

While soils in the tropical region showed K_d (alachlor) = 1.47 to 1.58 L kg^{-1} and K_{oc} (alachlor) = 65.10 to 80.90 L kg^{-1}, K_d (bentazone) = 0.18 to 0.37 L kg^{-1} and K_{oc} (bentazone) = 7.00 to 15.10 L kg^{-1}, K_d (imazethapyr) = 0.32 to 2.19 L kg^{-1} and K_{oc} (imazethapyr) = 8.70 to 95.90 L kg^{-1} (Bonfleur et al. 2016). In general, the K_d values were low in all soils (K_d <5.0 L kg^{-1}). The study also considered the effect of the size of the aggregates of each soil on the sorption of herbicides. Tropical soils showed greater sorption potential than subtropical soils considering the fraction of soil aggregates of 2–53 µm in size, which contributed to the sorption of 40%–56% and 44%–87% of herbicides in each region. The sorption percentage of the three herbicides correlated positively SOC and the clay content of the soils. This demonstrates the important role of SOC and soil mineralogy, as well as the interaction of these two factors in herbicide sorption.

Similarly, Toniêto et al. (2016) evaluated the sorption of ^{14}C-hexazinone and ^{14}C-tebuthiuron as a function of the amount of sugarcane straw in two types of soil. In addition, the authors also examined the aging of the herbicide in the soil. This study reported low apparent sorption coefficients ($K_{d\ app}$) (when using a single dose of the herbicide) for ^{14}C-tebuthiuron and ^{14}C-hexazinone, with $K_{d\ app}$ values of 3.25 and 1.76 L kg^{-1}, respectively. The clay Oxisol sorbed 41 and 56% of applied ^{14}C-hexazinone and ^{14}C-tebuthiuron, while the sandy clay loam sorbed 14% and 24% of ^{14}C-hexazinone and ^{14}C-tebuthiuron applied. Sugarcane straw promoted a slight increase in the sorption of both herbicides. However, the increase in sorption was greater for ^{14}C-tebuthiuron in sandy soil compared to control (without straw). The increase in the amount sorbed was 15%–24% for ^{14}C-tebuthiuron and 9%–12% for ^{14}C-hexazinone in soils recently managed with 10 Mg ha^{-1} of sugarcane straw. The straw aging also contributed to an increase in ^{14}C-tebuthiuron sorption (28%–30% at 21 days after application) but promoted a slight decrease in ^{14}C-hexazinone sorption, which decreased from 18% to 17%. The authors state that the contribution of the newly added straw had little effect on the sorption of both herbicides. On the other hand, the aging of the straw in the soil contributed to the increase in herbicide sorption. This is due to the decomposition of the straw and an increase in SOC content, confirming the increase in the sorption of both herbicides in the sandy soil. It is worth mentioning that both herbicides have a high potential for leaching ($K_{d\,app}$ <5.0 L kg^{-1}); however the management of sugarcane straw in the soil can reduce the risk of leaching of these herbicides.

The sorption of herbicides can also be affected by the application of soil conditioners, such as animal manure, biochar and biosolids. In this context, Junqueira et al. (2019) examined how biochar-amended soil and biochar aging affected the behavior of ^{14}C-glyphosate in two different soil classes (Ultisol-Typic Hapludalf and Alfisol-Paleudult). In Ultisol-Typic Hapludalf soil, the sorption coefficient was slight lower when comparing recently applied biochar (K_d = 17.57 L kg^{-1}) and aged biochar (K_d = 15.35 L kg^{-1}). On the other hand, in Alfisol-Paleudult soil the K_d values were practically the same for both fresh and aged biochar soil. In both soils, the application of biochar promoted an increase in K_d in relation to unamended soil. The authors also found a low desorption rate of ^{14}C-glyphosate, with values of 3.3%–3.4% in Ultisol-Typic Hapludalf soil and 3.9%–4.1% in Alfisol-Paleudult soil.

Evaluating the sorption and desorption of ^{14}C-aminocyclopyrachlor and ^{14}C-mesotrione as a function of the application of three different doses of municipal sewage sludge (SS) in the soil (1,2, 12 and 120 t ha^{-1}, dry basis), Mendes et al. (2019b) reported a significant increase in the Freundlich sorption coefficient (K_f) in response to increasing doses of sewage sludge. K_f values ranged from 1.07 to 1.45 μmol$^{(1-1/n)}$ L$^{1/n}$ kg^{-1} for ^{14}C-aminocyclopyrachlor and 3.48 to 4.25 μmol$^{(1-1/n)}$ L$^{1/n}$ kg^{-1} for ^{14}C-mesotrione.

The dose of 120 t ha^{-1} of SS promoted greater sorption for both herbicides (Mendes et al. 2019b). This fact was attributed to the increase in SOC due to the greater contribution provided by the higher dose of SS. However, the authors point out that ^{14}C-aminocyclopyrachlor (pK_a = 4.6) and ^{14}C-mesotrione (pK_a = 3.12) are weak acid herbicides, so the increase in pH decreases the sorption of these herbicides, since the application sludge increased the pH of the soil by 0.6 units, affecting the assorted behavior of herbicides. The desorption equations fitted the Freundlich model, with a high coefficient of determination [R^2 (aminocyclopyrachlor) = 0.99 and R^2 (mesotrione) = 0.97]. For ^{14}C-aminocyclopyrachlor, the values $1/n$ were ~1 for both sorption and desorption. While, the ^{14}C-mesotrione showed values of $1/n_{desorption}$ were 0.38–0.44 lower than the values of $1/n_{sorption}$, suggesting that the desorption of this herbicide is dependent on the concentration in the soil solution.

From the approach of precision agriculture with application of varied rate of herbicide for the control of weeds in pre-emergence, Mendes et al. (2020a) analyzed the interpolation of physical-chemical attributes of the soil with the retention of the herbicides ^{14}C-hexazinone and ^{14}C-tebuthiuron and use the soil/herbicide retention maps (Figure 2.9) as a tool to assist in the application, in the efficiency of the products and in reducing risks to the environment. Sorption data were widely variable between different collection points within the studied area. For sorption, the K_d values of ^{14}C-hexazinone and ^{14}C-tebuthiuron ranged from 1.21 to 2.92 L kg^{-1} and 0.35 to 0.63 L kg^{-1}, respectively. For desorption, K_d values ranged from 3.36 to 4.38 L kg^{-1} (^{14}C-hexazinone) and 2.56 to

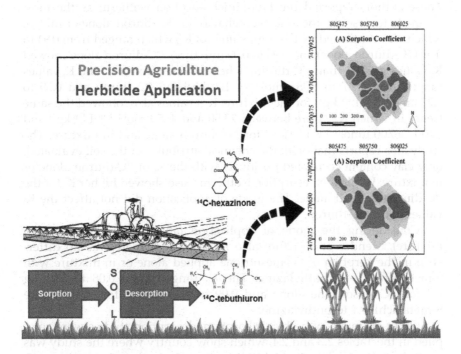

Figure 2.9 Interpolated maps of ^{14}C-hexazinone and ^{14}C-tebuthiuron sorption coefficient for precision agriculture (herbicide application) in an agricultural area (18.21 ha) in the municipality of Anhumas, São Paulo, Brazil. (Adapted from Mendes et al. 2020a.)

2.98 L kg^{-1} (^{14}C-tebuthiuron). Positive and strong Pearson correlations were observed between the sorption–desorption coefficients of both herbicides with the clay and SOM contents. This corroborates the fact that there is a greater interaction of these herbicides with the mineral fraction of the soil and SOM. In addition, the bioavailability of herbicides along the spatial gradient was also evaluated. In general, ^{14}C-hexazinone showed greater bioavailability than ^{14}C-tebuthiuron, with the bioavailable amounts after mass balance above 85% for ^{14}C-hexazinone and 45% for ^{14}C-tebuthiuron.

The soil depth effect on sorption–desorption of ^{14}C-hexazinone was studied by Chitolina et al. (2020), who also reported low sorption coefficient values in the 0-30 cm layer, the K_d values (0-10 cm layer) = 0.20 L kg^{-1}, K_d (10–20 cm layer) = 0.13 L kg^{-1} and K_d (20–30 cm layer) = 0.08 L kg^{-1}. In this study, the K_d values of ^{14}C-hexazinone gradually decreased with increasing depth and reducing of the SOM content.

Mixing products is also a practice that can affect the herbicide sorption behavior. Takeshita et al. (2019b) studied the sorption–desorption isotherms of ^{14}C-diuron alone and the ^{14}C-diuron + hexazinone mixture. These authors reported the Freundlich sorption isotherm as the most adequate to describe the assorted behavior of ^{14}C-diuron alone or mixed, which was confirmed by the high values of R^2, which ranged from 0.94 to 0.99 (^{14}C-diuron alone) and 0.95 to 0.99 (mixture). ^{14}C-diuron alone showed K_f values greater than ^{14}C-diuron + hexazinone (mixture). The K_f values ranged from 1.47 to 5.08 μmol$^{(1-1/n)}$ L$^{1/n}$ kg^{-1} for ^{14}C-diuron and 0.59 to 3.77 μmol$^{(1-1/n)}$ L$^{1/n}$ kg^{-1} for the mixture. K_{foc} values also followed the same trend, in which K_{foc} were between 73.50 and 445 μmol$^{(1-1/n)}$ L$^{1/n}$ kg^{-1} and 29.00–367.00 μmol$^{(1-1/n)}$ L$^{1/n}$ kg^{-1} for ^{14}C-diuron alone and in mixture. The study also reports that of all the chemical attributes of the soil evaluated, only clay content correlated positively with the K_f of ^{14}C-diuron alone or in mixture. While the desorption isotherms also showed higher K_f for the ^{14}C-diuron applied alone. The mode of application did not affect the K_f values of the ^{14}C-diuron.

Considering the mode of application (application alone or in a mixture), Mendes et al. (2016) also reported that there was no difference in the sorption of ^{14}C-mesotrione applied alone or in mixture with S-metalachlor + terbuthylazine. The K_d ranged from 0.08 to 5.05 kg L^{-1} for ^{14}C-mesotrione alone and 0.09 to 5.20 kg L^{-1} for the mixture S-metalachlor + terbuthylazine.

Other examples of herbicide sorption–desorption studies were compiled in the Tables 2.3 and 2.4, which show country where the study was carried out, name of the herbicide, the adjusted sorption–desorption coefficient of the isotherm model – linear (K_d), Freundlich (K_f) and Langmuir (K_l), sorption coefficient normalized by the organic carbon content by three models (K_{oc}, K_{foc}, K_{loc}), and references.

Table 2.3 Sorption coefficient and isotherm parameter of ^{14}C-herbicides in soils of the worldwide

Local (Country)	Soil Classification[a]	Soil Texture[b]	Radiolabeled Herbicide	K_d	K_{oc}	K_f	K_{foc}	1/n	R²	K_l	K_{loc}	q Max	R²	Sorption (%)	References
				L kg⁻¹		μmol$^{(1-1/n)}$ L$^{1/n}$ kg⁻¹				kg L⁻¹		g kg⁻¹			
Brazil	Terra Preta de Índio (anthropogenic soils)1	Clay Loam	^{14}C-diuron (phenyl-^{14}C (U))	28.34	761.83	8.74	234.90	0.70	0.99	0.01	0.37	0.84	0.99	99.10	Almeida et al. (2020)
	Terra Preta de Índio (anthropogenic soils)2	Clay Loam		21.93	565.21	6.38	164.40	0.66	0.99	0.01	0.36	0.75	0.99	98.90	
	Oxisol – Typic quartzipsaments[b]	Sand		0.50	151.52	0.41	124.20	0.71	0.99	0.36	110.30	0.37	0.99	60.90	
Brazil	Oxisol – Rhodic Hapludox[b]	Clay Loam	^{14}C-indaziflam (triazine-2,4-^{14}C)	21.14	1321.00	19.10	1194.00	0.95	0.99	-	-	-	-	90.00	Alonso et al. (2011)
	Oxisol – Rhodic Hapludox[b]	Clay Loam		27.44	1339.00	29.27	1428.00	1.03	0.99					91.00	
	Oxisol – Rhodic Hapludox[b]	Clay Loam		9.42	434.00	9.00	415.00	0.98	0.99					91.00	
	Oxisol – Typic Hapludox[b]	Loamy Sand		4.86	972.00	4.66	933.00	0.95	0.99					66.00	
	Oxisol – Arenic albaqualfs[b]	Sandy Loam		12.44	1173.00	11.87	1120.00	0.98	0.99					82.00	
	Oxisol – Typic quartzipsaments[b]	Sand		5.22	885.00	5.23	858.00	0.99	0.99					66.00	
USA	Mollisol – Typic calciaquolls[b]	Sandy Loam		12.96	595.00	10.86	498.00	0.95	0.99					84.00	
	Mollisol – Typic calciudolls[b]	Sandy Loam		7.17	652.00	6.62	602.00	0.94	0.99					75.00	
	Mollisol – Typic hapludolls[b]	Sandy Loam		16.72	664.00	14.26	566.00	0.92	0.99					87.00	

(Continued)

Table 2.3 Sorption coefficient and isotherm parameter of [14C]-herbicides in soils of the worldwide (*Continued*)

Local (Country)	Soil Classification[a]	Soil Texture[b]	Radiolabeled Herbicide	K_d	K_{oc} (L kg^{-1})	K_f	K_{foc} (μmol$^{(1-1/n)}$ L$^{1/n}$ kg^{-1})	$1/n$	R^2	K_l (kg L^{-1})	K_{loc}	q Max (g kg^{-1})	R^2	Sorption (%)	References
Brazil	Ultisol[b,d]	Sandy Loam	[triazine-6-[14C]]-hexazinone	-	-	0.12	23.10	0.75	0.97	-	-	-	-	14.30	Mendes et al. (2019a)
	Ultisol[b,e]					20.76	1958.90	0.72	0.99					99.10	
	Biochar					368.64	3771.20	0.78	0.99					99.90	
	Ultisol[b,d]		[ring-6-[14C]]-metribuzin			0.17	32.69	0.69	0.99					22.70	
	Ultisol[b,e]					29.31	2758.90	0.82	0.99					98.80	
	Biochar					149.43	1348.50	0.96	0.99					99.00	
	Ultisol[b,d]		[2,3,4-[14C]]-quinclorac			0.37	71.15	0.64	0.97					39.70	
	Ultisol[b,e]					467.28	40964.20	0.65	0.97					99.90	
	Biochar					485.35	4253.50	0.69	0.98					99.80	
Spain	Soil 1[c]	Silt Loamy	[ring-U-[14C]]-diuron	-		13.00	664.00	0.81	0.93	-		-	-	-	Rubio-Bellido et al. (2016)
	Soil 2[c]	Silt Loamy				13.80	1112.00	0.66	0.91						
	Soil 3[c]	Clay				13.20	877.00	0.54	0.99						
	Soil 4[c]	Sand Loamy			-	6.49	638.00	0.67	0.91						
	Soil 5[c]	Loamy				2.90	320.00	1.04	0.97						
	Soil 6[c]	Silt Loamy				14.30	1738.00	0.70	0.95						
	Soil 7[c]	Sand				0.60	516.00	0.90	0.95						

(*Continued*)

Table 2.3 Sorption coefficient and isotherm parameter of ^{14}C-herbicides in soils of the worldwide (*Continued*)

Local (Country)	Soil Classification[a]	Soil Texture[b]	Radiolabeled Herbicide	K_d (L kg⁻¹)	K_{oc} (L kg⁻¹)	K_f (μmol$^{(1-1/m)}$ L$^{1/n}$ kg⁻¹)	K_{foc}	1/n	R^2	K_l (kg L⁻¹)	K_{foc} / q Max (g kg⁻¹)	R^2	Sorption (%)	References
Brazil	Oxisol – Rhodic Hapludox[b]	Clay	^{14}C-aminocyclopyrachlor	0.62	39.00	0.63	39.00	0.97	0.99	-	-	-	-	Oliveira Jr. et al. (2011)
	Ultisol[b]	Sand		0.23	32.00	0.22	32.00	0.94	0.99					
	Oxisol – Typic Hapludox[b]	Loamy Sand		0.05	10.00	0.06	11.00	0.90	0.99					
	Oxisol – Typic Hapludox[b]	Loamy Sand		0.34	33.00	0.34	33.00	0.95	0.99					
	Oxisol – Rhodic Hapludox[b]	Clay		0.50	27.00	0.52	28.00	0.97	0.99					
	Plintosol[b]	Sandy Loam		0.29	32.00	0.29	32.00	0.99	0.99					
	Oxisol – Rhodic Hapludox[b]	Clay		1.17	57.00	1.16	56.00	1.01	0.98					
	Oxisol – Rhodic Hapludox[b]	Clay		0.85	39.00	0.83	38.00	1.00	0.99					
	Typic quartzipsaments[b]	Sand		0.06	9.00	0.07	11.00	0.93	0.99					
	Oxisol – Typic Hapludox[b]	Clay		0.48	27.00	0.49	27.00	0.98	0.99					
	Arenic albaqualfs[b]	Sandy Loam		0.50	47.00	0.50	47.00	0.99	0.99					
	Typic quartzipsaments[b]	Sand		0.09	16.00	0.09	15.00	0.97	0.99					
	Oxisol – Typic Hapludox[b]	Clay		1.07	66.00	1.05	64.00	0.99	0.99					
	Oxisol – Typic Hapludox[b]	Loamy Sand		0.14	22.00	0.14	22.00	1.03	0.99					

(*Continued*)

Table 2.3 Sorption coefficient and isotherm parameter of ^{14}C-herbicides in soils of the worldwide (*Continued*)

Local (Country)	Soil Classification[a]	Soil Texture[b]	Radiolabeled Herbicide	K_d (L kg^{-1})	K_{oc} (L kg^{-1})	K_f (μmol$^{(1-1/n)}$ L$^{1/n}$ kg^{-1})	K_{foc} (μmol$^{(1-1/n)}$ L$^{1/n}$ kg^{-1})	$1/n$	R^2	K_l (kg L^{-1})	K_{loc} (kg L^{-1})	q Max (g kg^{-1})	R^2	Sorption (%)	References
USA	Farming soil[c]	Sand Loamy	[ring-U-14-clomazone]	11.00	887.00	-	-	-	-	-	-	-	-	-	Gunasekara et al. (2009)
	Farming soil[c]	Sandy Clay Loam		5.50	447.00										
	Forest soil[c]	Clay		10.00	378.00										
	Forest soil[c]	Clay		7.00	280.00										
Malaysia	Bernam[c]	Loamy Clay	[triazine-2-14C] of metsulfuron-methyl	-	-	2.84	-	0.87	-	-	-	-	-	-	Ismail and Ooi (2012)
	Selangor[c]	Loamy Clay				1.57		0.91							
	Rengam[c]	Sandy Loam				0.56		0.87							
	Bongor[c]	Sandy Loam				0.37		0.87							
Bangladesh	Silmondy site[c]	Clay	[14C-phenyl ring]-oxadiazon	4.98	530.30	-	-	-	-	-	-	-	-	-	Hoque et al. (2007)
	Sonatala site[c]	Silt Clay		7.04	757.20										
United Kingdom	Hallsworth site[c]	Sandy Clay		9.37	385.8										
Colombia	Colombian soil[c]	Silt Clay	[Ring-U-14C]-diuron	10.3	1118	-	-	-	-	-	-	-	-	84.3	Peña-Martínez et al. (2018)
Spain	Spain soil[c]	Loam		20.7	1903									91.3	
Colombia	Colombian soil[c]	Silt Clay	[Ring-14C]-ametryn	8.5	929									81.2	
Spain	Spain soil[c]	Loam		3.6	327									64.4	

[a] According to Soil Classification WRB/FAO
[b] According to American Soil Taxonomy
[c] n.a., non-available
[d] Us, unamended soil
[e] Asb, amended soil with biochar

Table 2.4 Desorption coefficient and isotherm parameter of ^{14}C-herbicides in soils of the worldwide

Local (Country)	Soil Classification[1/]	Soil texture	Radiolabeled Herbicide	K_d L kg^{-1}	K_{oc} L kg^{-1}	K_f μmol$^{(1-1/n)}$ L$^{1/n}$ kg^{-1}	K_{foc}	1/n	R^2	K_l kg L^{-1}	K_{loc}	q Max g kg^{-1}	R^2	Desorption (%)	References
Brazil	Terra Preta de Índio (anthropogenic soils)1	Clay loam	^{14}C-diuron (phenyl-^{14}C (U))	132.00	3548.40	50.40	1355.11	0.82	0.99	0.008	0.22	1.64	0.99	1.36	Almeida et al. (2020)
	Terra Preta de Índio (anthropogenic soils)2	Clay loam		74.17	1911.60	13.50	347.94	0.65	0.99	0.004	0.10	0.76	0.99	1.70	
	Oxisol – Typic quartzipsaments[b]	Sand		3.55	1075.00	3.05	924.24	0.93	0.99	0.382	115.76	1.57	0.99	24.02	
Brazil	Oxisol – Rhodic Hapludox[b]	Clay loam	[triazine-2,4-^{14}C]-indaziflam	39.19	-	-	-	0.04	0.65	-	-	-	-	90.00	Alonso et al. (2011)
	Oxisol – Rhodic Hapludox[b]	Clay loam		7.04				0.29	0.82					91.00	
	Oxisol – Rhodic Hapludox[b]	Clay loam		38.95				0.09	0.86					91.00	
	Oxisol – Typic Hapludox[b]	Loamy sand		15.80				0.11	0.80					66.00	
	Oxisol – Arenic albaqualfs[b]	Sandy loam		16.18				0.25	0.95					82.00	
	Oxisol – Typic quartzipsaments[b]	Sand		8.18				0.24	0.94					66.00	
USA	Mollisol – Typic calciaquolls[b]	Sandy loam		19.60				0.13	0.86					84.00	
	Mollisol – Typic calciudolls[b]	Sandy loam		9.98				0.28	0.82					75.00	
	Mollisol – Typic hapludolls[b]	Sandy loam		23.52				0.16	0.95					87.00	

(Continued)

Table 2.4 Desorption coefficient and isotherm parameter of ^{14}C-herbicides in soils of the worldwide *(Continued)*

Local (Country)	Soil Classification[1/]	Soil texture	Radiolabeled Herbicide	K_d L kg⁻¹	K_{oc}	K_f µmol$^{(1-1/n)}$ L$^{1/n}$ kg⁻¹	K_{foc}	$1/n$	R^2	K_l kg L⁻¹	K_{loc} g kg⁻¹	q Max	R^2	Desorption (%)	References
Brazil	Ultisol[b,d]	Sandy loam	[triazine-6-^{14}C]-hexazinone	-	-	17.96	3150.96	1.19	0.97	-	-	-	-	6.58	Mendes et al. (2019a)
	Ultisol[b,e]					57.89	5392.99	0.78	0.99					0.55	
	Biochar					1125.00	10164.18	0.84	0.98					0.06	
	Ultisol[b,d]		[ring-6-^{14}C]-metribuzin			17.93	3539.42	1.28	0.99					9.82	
	Ultisol[b,e]					103.83	9842.05	0.83	0.99					0.39	
	Biochar					1088.10	9680.36	1.07	0.99					0.44	
	Ultisol[b,d]		[2,3,4-^{14}C]-quinclorac			6.60	1274.03	1.04	0.96					12.12	
	Ultisol[b,e]					1253.50	111692.99	0.72	0.96					0.42	
	Biochar					1302.82	11855.18	0.75	0.97					0.05	
Spain	Soil 1[c]	Silt loamy	[ring-U-^{14}C]-diuron	-	n.a.	13.00	-	-	-	-	-	-	-	15.90	Rubio-Bellido et al. (2016)
	Soil 2[c]	Silt loamy				13.80								20.10	
	Soil 3[c]	Clay				13.20								11.20	
	Soil 4[c]	Sand loamy				6.49								29.50	
	Soil 5[c]	Loamy				2.90								43.30	
	Soil 6[c]	Silt loamy				14.30								10.70	
	Soil 7[c]	Sand				0.60								73.70	

(Continued)

Table 2.4 Desorption coefficient and isotherm parameter of ^{14}C-herbicides in soils of the worldwide *(Continued)*

Local (Country)	Soil Classification[1]/	Soil texture	Radiolabeled Herbicide	K_d (L kg⁻¹)	K_{oc} (L kg⁻¹)	K_f (μmol$^{(1-1/n)}$ L$^{1/n}$ kg⁻¹)	K_{foc}	1/n	R^2	K_l (kg L⁻¹)	K_{loc}	q Max (g kg⁻¹)	R^2	Desorption (%)	References
Brazil	Oxisol – Rhodic Hapludox[b]	Clay	^{14}C-aminocyclopyrachlor	1.48	-	-	-	0.21	0.97	-	-	-	-	-	Oliveira Jr. et al. (2011)
	Ultisol[b]	Sand		0.48				0.24	0.71						
	Oxisol – Typic Hapludox[b]	Loamy sand		0.12				0.24	0.83						
	Oxisol – Typic Hapludox[b]	Loamy sand		0.60				0.33	0.96						
	Oxisol – Rhodic Hapludox[b]	Clay		0.93				0.25	0.94						
	Plintosol[b]	Sandy loam		0.38				0.65	0.99						
	Oxisol – Rhodic Hapludox[b]	Clay		1.71				0.38	0.90						
	Oxisol – Rhodic Hapludox[b]	Clay		1.28				0.41	0.99						
	Typic quartzipsaments[b]	Sand		0.13				0.29	0.73						
	Oxisol – Typic Hapludox[b]	Clay		0.76				0.44	0.98						
	Arenic albaqualfs[b]	Sandy loam		0.75				0.45	0.94						
	Typic quartzipsaments[b]	Sand		0.19				0.30	0.96						
	Oxisol – Typic Hapludox[b]	Clay		1.46				0.47	0.97						
	Oxisol – Typic Hapludox[b]	Loamy sand		0.21				0.63	0.97						

(Continued)

Table 2.4 Desorption coefficient and isotherm parameter of ^{14}C-herbicides in soils of the worldwide *(Continued)*

Local (Country)	Soil Classification[1/]	Soil texture	Radiolabeled Herbicide	K_d	K_{oc}	K_f	K_{foc}	1/n	R^2	K_l	K_{loc} q Max kg L⁻¹	g kg⁻¹	R^2	Desorption (%)	References
				L kg⁻¹		µmol $^{(1-1/n)}$ L$^{1/n}$ kg⁻¹									
Malaysia	Bernam[c]	Loamy clay	[triazine-2-¹⁴C] of metsulfuron-methyl	-	-	2.56	-	0.28	-	-	-	-	-	-	Ismail and Ooi (2012)
	Selangor[c]	Loamy clay				1.41		0.44							
	Rengam[c]	Sandy loam				1.06		0.33							
	Bongor[c]	Sandy loam				0.56		0.52							

[a] According to Soil Classification WRB/FAO
[b] According to American Soil Taxonomy
[c] n.a., non-available
[d] Us, unamended soil
[e] Asb, amended soil with biochar

2.8 Concluding remarks

The study of the behavior of herbicides in soils is already consolidated in the scientific community, and the use of radiolabeled herbicides for these studies has been increasing in use and importance. The plant–soil–herbicide interaction alters the availability of the herbicide in the soil solution and is governed mainly by the processes of retention in colloidal particles of the soil and absorption by weeds. The methodologies that involve these practices are still relatively widespread in the literature.

In this chapter, detailed methodologies for sorption and desorption studies of radiolabeled herbicides in soils were reported, because the use of theses herbicides promote scientific advantages, such as the possibility of accurately determining minimum amounts of the products in relatively short periods and their fate in the soil.

The studies described in this chapter contemplate the sorption and desorption processes, using the batch equilibrium and stirred flow method. In addition to the description of the study methods, additional information is addressed, such as, for example, the characteristics of the radiolabeled products and information from previous studies according to OECD and EPA guidelines. This information is of great relevance for every researcher who wishes to use radiolabeled herbicides in their scientific research aimed at studying sorption and desorption.

References

Almeida, C. S., Mendes K. F., Junqueira L. V., Alonso F. G., Chitolina G. M. and Tornisielo V. L. 2020. Diuron sorption, desorption and degradation in anthropogenic soils compared to sandy soil. *Planta Daninha* 38:1–14.

Ayawei, N., Ebelegi A. N. and Wankasi D. 2017. Modelling and interpretation of adsorption isotherms. *Journal of Chemistry* 2017:1–12.

Alonso, D. G., Koskinen W. C., Oliveira R. S., Constantin J. and Mislankar S. 2011. Sorption–desorption of Indaziflam in selected agricultural soils. *Journal of Agricultural and Food Chemistry* 59(24):13096–13101.

Amézqueta, S., Subirats X., Fuguet E., Rosés M. and Ráfols C. 2005. Octanol-water partition constant. In: *Liquid-Phase Extraction* ed. C. F. Poole, 138–208. Amsterdam: Elsevier Inc.

Amondham, W., Parkpian P., Polprasert C., Delaune R. D. and Jugsujinda A. 2006. Paraquat adsorption, degradation, and remobilization in tropical soils of Thailand. *Journal of Environmental Science and Health, Part B* 41:485–507.

Ball, W.P. and Roberts P.V. 1991. Long-term sorption of halogenated organic chemicals by aquifer material: 1 equilibrium. *Environmental Science and Technology* 25(7):1223–1236.

Barizon, R. R. M., Lavorenti A., Regitano J. B. and Tornisielo V. L. 2005. Imazaquin sorption and desorption in soils with different mineralogical, physical and chemical characteristics. *Revista Brasileira de Ciência do Solo* 29:695–703.

Barriuso, E., Laird D. A., Koskinen W. C. and Dowdy R. H. 1994. Atrazine desorption from smectites. *Soil Science Society of America Journal* 58(6):1632–1638.

Barrow, N. J. 1979. The description of desorption of phosphate from soil. *Journal of Soil Science* 30:259–270.

Berkowitz, B., Dror I. and Yaron B. 2008. *Contaminant Geochemistry: Interactions and Transport in the Subsurface Environment*. Berlin: Elsevier.

Boesten, J. J. T. I. 2016. Proposal for field-based definition of soil bound pesticide residues. *Science of the Total Environment* 544:114–117.

Bonfleur, E. J., Kookana R. S., Tornisielo V. L. and Regitano J. B. 2016. Organomineral interactions and herbicide sorption in Brazilian tropical and subtropical oxisols under no-tillage. *Journal of Agricultural and Food Chemistry* 64(20):3925–3934.

Braun, K. E., Luks A. K. and Schmidt B. 2017. Fate of the ^{14}C-labeled herbicide prosulfocarb in a soil and in a sediment-water system. *Journal of Environmental Science and Health, Part B* 52:122–130.

Burwell, R. L. 1976. *Manual of Symbols and Terminology for Physicochemical Quantities and Units. Appendix 2: Definitions, Terminology and Symbols in Colloid and Surface Chemistry. Part 2, Heterogeneous Catalysis*, 1st ed. Amsterdam: Elsevier.

Celis, R. and Koskinen W. C. 1999. An isotopic exchange method for the characterization of the irreversibility of pesticide sorption–desorption in soil. *Journal of Agricultural and Food Chemistry* 47:782–90.

Chi, Y., Zhang G., Xiang Y., Cai D. and Wu Z. 2017. Fabrication of a temperature-controlled-release herbicide using a nanocomposite. *Sustainable Chemistry & Engineering* 5(6):4969–4975.

Chiou, C. T. 2002. *Partition and Adsorption of Organic Contaminants in Environmental Systems*. Hoboken: Wiley-Interscience.

Chitolina, G. M., Mendes K. F., Almeida C. S., Alonso F. G., Junqueira L. V. and Tornisielo V. L. 2020. Influence of soil depth on sorption and desorption processes of hexazinone. *Planta Daninha* 38:1–8.

Conde-Cid, M., Paradelo R., Fernández-Calviño D., Pérez-Novo C., Nóvoa-Múñoz J. C. and Arias-Estévez M. 2017. Retention of quaternary ammonium herbicides by acid vineyard soils with different organic matter and Cu contents. *Geoderma* 293:26–33.

Correia, F. V., Macrae A., Guilherme L. R. G. and Langenbach T. 2007. Atrazine sorption and fate in a Ultisol from humid tropical Brazil. *Chemosphere* 67:847–854.

Environmental Protect Agency (EPA). 1999. Understanding variation in Partition coefficient, Kd, values. https://www.epa.gov/radiation/understanding-variation-partition-coefficient-kd-values (accessed May 14, 2020).

Environmental Protect Agency (EPA). 2008. Series 385 - Fate, transport and transformation test guidelines. OPPTS 835.1210. https://www.epa.gov/test-guidelines-pesticides-and-toxic-substances/series-835-fate-transport-and-transformation-test (accessed May 27, 2020).

Gavrilescu, M. 2005. Fate of pesticides in the environment and its bioremediation. *Engineering in Life Sciences* 5(6):497–526.

Gevao, B., Semple K. T. and Jones K. C. 2000. Bound pesticide residues in soils: A review. *Environmental Pollution* 108:3–14.

Giles, C., Smith D. and Huitson A. 1974. A general treatment and classification of the solute adsorption isotherm. I. Theoretical. *Journal of Colloid and Interface Science* 47(3):755–765.

Gill, H. K. and Garg H. 2014. Pesticides: Environmental impacts and management strategies. In: *Pesticides: Toxic Aspects*, ed. M. L. Larramendy and S. Soloneski, 187–230. Rijeka: InTech.

Green, R. E. and Karickhoff S. W. 1990. *Pesticides in the Soil Environment: Processes, Impacts, and Modeling.* Madison, WI, USA: Soil Science Society of America

Gunasekara, A. S., dela Cruz I. D. P., Curtis M. J., Claassen V. P. and Tjeerdema R. S. 2009. The behavior of clomazone in the soil environment. *Pest Management Science* 65:711–716.

Helling, C. S. and Turner B. C. 1968. Pesticide mobility: Determination by soil thin-layer chromatography. *Science* 162:562–563.

Herwig, U., Narres H. D. and Schwuger M. J. 2001. Physicochemical interactions between atrazine and clay minerals. *Applied Clay Science* 18(1):201–22.

Heyse, E., Dai D., Rao P. S. C. and Delfino J. J. 1997. Development of a continuously stirred flow cell for investigating sorption mass transfer. *Journal of Contaminant Hydrology* 25:337–355.

Hoque, M. E., Wilkins R. M., Kennedy A. and Garratt J. A. 2007. Sorption behaviour of oxadiazon in tropical rice soils. *Water Science and Technology* 56:115–121.

Huang, W., Peng P., Yu Z. and Fu J. 2003. Effects of organic matter heterogeneity on sorption and desorption of organic contaminants by soils and sediments. *Applied Geochemistry* 18(7):955–972.

Ismail, B. S. and Ooi K. E. 2012. Adsorption, desorption and mobility of metsulfuron-methyl in soils of the oil palm agroecosystem in Malaysia. *Journal of Environmental Biology* 33:573–577.

IUSS Working Group WRB. 2014. *World Reference Base for Soil Resources.* Rome: FAO.

Junqueira, L. V., Mendes K. F., Sousa R. N., Almeida C. S., Alonso F. G. and Tornisielo V. L. 2019. Sorption–desorption isotherms and biodegradation of glyphosate in two tropical soils aged with eucalyptus biochar. *Archives of Agronomy and Soil Science* 1:1–17.

Karickhoff, S.W. and Morris K.R. 1985. Sorption dynamics of hydrophobic pollutants in sediment suspensions. *Environmental Toxicology and Chemistry* 4(4):469–479.

Kniss, A. R., Vassios J. D., Nissen S. J. and Ritz C. 2011. Nonlinear regression analysis of herbicide absorption studies. *Weed Science* 59:601–10.

Koskinen, W. C. and Harper S. S. 1990. The retention process: Mechanism. In: *Pesticides in the Soil Environmental: Processes, Impacts and Modeling* ed. H. H. Cheng, 51–77. Madison: Soil Science Society of America, Madison.

Koskinen, W. C., Ochsner T. E., Stephens B. M. and Kookana R. S. 2006. Sorption of isoxaflutole diketonitrile degradate (DKN) and dicamba in unsaturated soil. *Journal of Environmental Science and Health* 41:1071–1083.

Lenheer, J. A. and Aldrichs J. 1971. A kinetic and equilibrium study of the adsorption of carbaryl and parathion upon soil organic matter surfaces. *Soil Science Society of America Proceedings* 35:700–705.

Limousin, G., Gaudet G. P., Charlet L., Szenknect S., Barthe V. and Krimissa M. 2007. Sorption isotherms: A review on physical bases, modeling and measurement. *Applied Geochemistry* 22(2):249–275.

Liu, Y., Xu Z., Wu X., Gui W. and Zhu G. 2010. Adsorption and desorption behavior of herbicide diuron on various Chinese cultivated soils. *Journal of Hazardous Materials* 178(1–3):462–468.

Mackay, D., Shiu W. Y. and Ma K. 1997. *Illustrated Handbook of Physical-Chemical Properties and Environmental Fate for Organic Chemicals.* Boca Raton: CRC Press/Taylor & Francis.

Mamy, L. and Barriuso E. 2007. Desorption and time-dependent sorption of herbicides in soils. *European Journal of Soil Science* 58:174–187.

Martinez, S. J. F., Pachepsky Y. A. and Rawls W. J. 2010. Modelling solute transport in soil columns using advective-dispersive equations with fractional spatial derivatives. *Advances in Engineering Software* 41:4–8.

Mendes, K. F., Martins B. A. B., Reis F. C., Dias A. C. R. and Tornisielo V. L. 2017. Methodologies to study the behavior of herbicides on plants and the soil using radioisotopes. *Planta Daninha* 35:1–21.

Mendes, K. F., Alonso F. G., Mertens T. B., Inoue M. I., Oliveira M. G. and Tornisielo V. L. 2019b. Aminocyclopyrachlor and mesotrione sorption–desorption in municipal sewage sludge-amended soil. *Bragantia* 78:131–140.

Mendes, K. F., Wei M. C. F., Furtado I. V., Takeshita V., Pissolito J. P., Molin J. P. and Tornisielo V. L. 2020a. Spatial distribution of sorption and desorption process of ^{14}C-labelled hexazinone and tebuthiuron in tropical soil. *Chemosphere* (*submitted manuscript*).

Mendes, K. F., Reis M. R., Dias A. C. R., Formiga J. A., Christoffoleti P. J. and Tornisielo V. L. 2014. A proposal to standardize herbicide sorption coefficients in Brazilian tropical soils compared to temperate soils. *International Journal of Food, Agriculture and Environment* 12:424–433.

Mendes, K. F., Reis M. R., Inoue M. H., Pimpinato R. F. and Tornisielo V. L. 2016. Sorption and desorption of mesotrione alone and mixed with S-metolachlor + terbuthylazine in Brazilian soils. *Geoderma* 280:22–28.

Mendes, K. F., Sousa R. N., Takeshita V., Alonso F. G., Régo A. P. J. and Tornisielo V. L. 2019a. Cow bone char as a sorbent to increase sorption and decrease mobility of hexazinone, metribuzin, and quinclorac in soil. *Geoderma* 343:40–49.

Mendes, K. F., Souza R. N., Goulart M. O. and Tornisielo V. L. 2020b. Role of raw feedstock and biochar amendments on sorption-desorption and leaching potential of three ^{3}H and ^{14}C-labelled pesticides in soils. *Journal of Radioanalytical and Nuclear Chemistry* 324:1373–1386.

Organisation for Economic Co-operation and Development (OECD). 2000. OECD Guidelines for the testing of chemicals, Section 1. Test No. 106: Adsorption – desorption using a batch equilibrium. https://www.oecd-ilibrary.org/environment/test-no-106-adsorption-desorption-using-a-batch-equilibrium-method_9789264069602-en (accessed May 27, 2020).

Oliveira Jr., R. S. and Regitano J. B. 2009. Dinâmica de Pesticidas no Solo. In: *Química e Mineralogia do Solo - Parte 2: Aplicações* eds. Melo, V. F. and L. R. F. Alleoni, 187–248, Viçosa: Sociedade Brasileira de Ciência do Solo.

Oliveira Jr., R. S., Alonso D. G. and Koskinen W. C. 2011. Sorption–desorption of aminocyclopyrachlor in selected Brazilian soils. *Journal of Agricultural and Food Chemistry* 59(8):4045–4050.

Oliveira Jr., R. S., Koskinen W. C. and Ferreira F. A. 2001. Sorption and leaching potential of herbicides on Brazilian soils. *Weed Research* 41(2):97–110.

Oliveira Jr., R. S., Koskinen W. C., Graff C. D., Anderson J. L., Mulla D. J., Nater E. A. and Alonso D. G. 2013. Acetochlor persistence in surface and subsurface soil samples. *Water, Air, & Soil Pollution* 224:1–9.

Oliveira, M. F., Prates H. T., Santanna D. P. and Oliveira Jr. R. S. 2006. Imazaquin sorption in surface and subsurface soil samples. *Pesquisa Agropecuária Brasileira* 41:461–468.

Oliveira, M. and Brighenti A. M. 2011. Comportamento dos herbicidas no ambiente. In: *Biologia e Manejo de Plantas Daninhas* ed. R. S. Oliveira Jr., J. Constantin and M. H. Inoue, 263–304. Curitiba: Omnipax.

Peña-Martínez, Y. R., Martínez M. J. and Guerrero-Dallos J. A. 2018. Adsorción-desorción de diurón y ametrina en suelos de Colombia y España. *Revista Colombiana de Química* 47: 31–40.

Peng, L., Liu P., Feng X., Wang Z., Cheng T., Liang Y., Lin Z. and Shi Z. 2018. Kinetics of heavy metal adsorption and desorption in soil: Developing a unified model based on chemical speciation. *Geochimica et Cosmochimica Acta* 224:282–300.

Pereira, G. A. M., Rodrigues D. A., Fonseca L. A. B. V., de Jesus Passos, A. B. R., da Silva, M. R. F., Silva, D. V., & da Silva, A. A. 2018. Sorption and desorption behavior of herbicide clomazone in soils from Brazil. *Bioscience Journal* 34(6):1496–1504.

Pignatello, J.J. 2000. The measurement and interpretation of sorption and desorption rates for organic compounds in soil media. *Advances in Agronomy* 69:1–73.

Rahman, M.M., Liedl R. and Grathwohl P. 2004. Sorption kinetics during macropore transport of organic contaminants in soils: Laboratory experiments and analytical modeling. *Water Resources Research* 40:1–11.

Rigotto, R. M., Vasconcelos D. P. and Rocha M. M. 2014. Pesticide use in Brazil and problems for public health. *Cadernos de Saúde Pública* 30:1360–1362.

Riise, G and Salbu B. 1992. Mobility of dichlorprop in the soil-water system as a function of different environmental factors. I. A batch experiment. *Science of the Total Environment* 123:399–409.

Rubio-Bellido, M., Morillo E. and Villaverde J. 2016. Effect of addition of HPBCD on diuron adsorption-desorption, transport and mineralization in soils with different properties. *Geoderma* 265:196–203.

Schäffer, A., Kästner M. and Trapp S. 2018. A unified approach for including non-extractable residues (NER) of chemicals and pesticides in the assessment of persistence. *Environmental Sciences Europe* 30(51):1–14.

Schwarzenbach, R. P., Gschwend P. M. and Imboden D.M. 1993. *Environmental Organic Chemistry*. New York: Wiley.

Senesi, N., Padovano G., Loffredo E. and Testini C. 1986. *Interaction of amitrole, alachlor, and cycloate with humic acids*. In: 2nd International Conference Environmental Contamination, Amsterdam. *Proceedings* 169–171.

Seybold, C.A. and Mersie W. 1996. Adsorption and desorption of atrazine, deethylatrazine, deisopropylatrazine, hydroxyatrazine, and metolachlor in two soils from Virginia. *Journal of Environmental Quality* 44(7):1925–1929.

Sharma, A. D. and Lai D. 2019. Sorption of radiolabelled glyphosate on biochar aged in contrasting soils. *Journal of Environmental Science and Health* 54:49–53.

Singh, B., Farenhorst A., Gaultier J., Pennock D., Degenhardt D. and McQueen R. 2014. Soil characteristics and herbicide sorption coefficients in 140 soil profiles of two irregular undulating to hummocky terrains of western Canada. *Geoderma* 232–234:107–116.

Spark, K. M. and Swift R. S. 2002. Effect of soil composition and dissolved organic matter on pesticide sorption. *The Science of the Total Environment* 298:147–161.

Sparks, D. L. 2003. Kinetics of Soil Chemical Processes. In: *Environmental Soil Chemistry* ed. D. L. Sparks, 207–244. Amsterdam: Elsevier Inc.

Sposito, G. 2016. *The Chemistry of Soils*. 3rd edition. Oxford: Oxford University Press.

Steinberg, S.M., Pignatello J. J. and Sawhney B. L. 1987. Persistence of 1,2-dibromoethane in soils: Entrapment in intraparticle micropores. *Environmental Science & Technology* 21(12):1201–1208.

Strawn, D. G. and Sparks D. L. 2000. Effects of soil organic matter on the kinetics and mechanisms of Pb(II) sorption and desorption in soil. *Soil Science Society of America Journal* 64:144–156.

Sun, W. and Selim H. M. 2019. A general stirred-flow model for time-dependent adsorption and desorption of heavy metal in soils. *Geoderma* 347:25–31.

Takeshita, V., Mendes K. F., Alonso F. G. and Tornisielo V. L. 2019a. Effect of organic matter on the behavior and control effectiveness of herbicides in soil. *Planta Daninha* 37: e019214401.

Takeshita, V., Mendes K. F., Christoffoleti P. J., Tornisielo V. L. and Guimarães A. C. D. 2019b. Sorption–desorption isotherms of diuron alone and in a mixture in soils with different physico-chemical properties. *African Journal of Agricultural Research* 14:672–679.

Tan, K. H. 2011. *Principles of Soil Chemistry*. Boca Raton: CRC Press/Taylor & Francis.

Toniêto, T. A. P., Pierri L., Tornisielo V. L. and Regitano J. B. 2016. Fate of tebuthiuron and hexazinone in green-cane harvesting system. *Journal of Agricultural and Food Chemistry* 64:3960–3966.

Tsai, W. T. and Chen H. R. 2013. Adsorption kinetics of herbicide paraquat in aqueous solution onto a low-cost adsorbent, swine-manure-derived biochar. *International Journal of Environmental Science and Technology* 10:1349–1356.

Tuin, B. J. W. and Tels M. 1991. Continuous treatment of heavy metal contaminated clay soils by extraction in stirred tanks and in a countercurrent column. *Environmental Technology* 12:179–190.

Wang, W., Wang Y., Li Z., Wang H., Yu Z., Lu L. and Ye Q. 2014. Studies on the anoxic dissipation and metabolism of pyribambenz propyl (ZJ0273) in soils using position-specific radiolabeling. *Science of the Total Environment* 472:582–589.

Wauchope, R. D., Yeh S., Linders B. H. J., Kloskowski R., Tanaka K., Rubin B., Katayama A., Kordel W., Gerstl Z., Lane M. and Unsworth J. B. 2002. Pesticide soil sorption parameters: Theory, measurement, uses, limitations and reliability. *Pest Management Science* 58(2):419–445.

Weber, J. B., McKinnon E. J. and Swain L. R. 2003. Sorption and mobility of ^{14}C-labeled imazaquin and metolachlor in four soils as influenced by soil properties. *Journal of Agricultural and Food Chemistry* 51:5752–5759.

Weber, J.B. and Miller C.T. 1989. Organic chemical movement over and through soil. In: *Reactions and Movement of Organic Chemicals in Soils* eds. Sawhney, B. L. and K. Brown, 305–335. Madison, WI, USA: Soil Science Society of America Special Publications 22.

Weber, J.B., Shea P.H. and Weed S.B. 1986. Fluridone retention and release in soils. *Soil Science Society of America Journal* 50(3):582–588.

Wu, S. and Gschwend P.M. 1986. Sorption kinetics of hydrophobic organic compounds to natural sediments and soils. *Environmental Science & Technology* 20(7):717–725.

chapter three

Mobility studies of herbicides in the soil: soil thin-layer chromatography, leaching columns, and lysimeters

Vanessa Takeshita[1], Kassio Ferreira Mendes[2],
Leonardo Vilela Junqueira[1], Felipe Gimenes Alonso[1],
Nicoli Gomes de Moraes[1], Valdemar Luiz Tornisielo[1]
[1]*Center of Nuclear Energy in Agriculture, University of São Paulo*
[2]*Department of Agronomy, Federal University of Viçosa*

Contents

3.0   Introduction

The application of herbicides is a common process in agriculture, as these molecules can follow several routes in the environment, such as absorption and translocation by plants and organisms, or dissipate in the environment. Once in the environment, herbicides can incur a risk to the entire agro-ecosystem (Luchini and Andréa 2018). Regardless of the application mode (pre-emergence or post-emergence of the weed and/or crop), a large amount of the herbicide has, as its final destination, the soil (Khan 1980; Pimentel 1995). However, for herbicides applied in pre-emergence, the risk of transport and dissipation in the environment may be even more pronounced.

Transport is one of the processes for the dissipation of herbicides in the environment that depends on the physical and chemical characteristics of the products, the soil, and the environmental conditions. The movement of herbicides in the soil depends on the flow of the water in the system, allowing these compounds to reach surface and groundwater. This movement occurs in different directions of the soil, through the diffusion and/or mass flow of water, associated with the luminous, chemical, and biological transformations and the processes of sorption and desorption of the product (Hunter and Stobbe 1972; Barriuso et al. 1992). Studies that characterize this process, and the relative importance of each factor interfering in the mobility of the herbicide in the soil, are fundamental in understanding the risks and sustainable chemical management of these products (Carter 2000).

Among the herbicide transport processes in the soil, leaching is described as one of the main processes, especially for non-volatile and water-soluble products (Monquero et al. 2010). The flow of water can move herbicides to areas with less absorption by weed roots (Inoue et al. 2010), thus being able to contaminate groundwater (Flury 1996). However, this process is also important in the control of weeds and management of the seed bank in the soil, where the herbicides must be leached to the region of the roots of the plants to be absorbed and reach the site of action (Oliveira 2001; Christoffoleti et al. 2008).

The main herbicide factors that affect their leaching are related to the physicochemical characteristics of the molecules, in addition to the formulation and additives (Oliveira and Brighenti 2011). An herbicide with high water solubility, for example, will be more easily dissolved in the soil solution, and may suffer greater leaching than an herbicide with low water solubility, which may be more absorbed in the soil colloids.

To estimate the mobility of herbicides in the soil, several simulations can be developed in the laboratory (soil columns and thin layer chromatography [TLC]) and in the field (lysimeter), in order to predict the behavior of each product. The use of radiometric techniques, with marked carbon molecules (^{14}C), associated with these simulations is essential for the registration of the herbicide in the country and provides information on the risk of leaching for each herbicide. This technique is an analytical tool of immense contribution in weed science, as it can provide high sensitivity, precision, specificity, and fast analysis, being a well-established tool for those types of approaches (Silva et al. 2009; Luchini and Andréa 2018). When combined with other conventional analytical techniques it can offer greater precision in measuring very small quantities in a short time (Luchini and Andréa 2018).

According Luchini and Andréa (2018), the application of radiometric methods for behavior of herbicides increases the capacity of laboratory studies. For tests of mobility of herbicides in the soil, guidelines that determine standards and protocols can be followed that indicate studies of soil TLC (EPA 1998), column leaching (OECD 2004, EPA 2008), and leaching studies in lysimeters (OECD 2000) for the characterization of herbicide mobility in the soil. In addition, there are other published works mentioned throughout the chapter that can help describe these methods.

This chapter describes the most relevant terms and methods used to characterize the mobility of herbicides in the soil using radiometric techniques, as well as an indication of how the data obtained through these studies should be interpreted and applied in the agricultural scenario of their use, along with management of weeds and the environmental risk of herbicides.

3.1 Important concepts for the study of herbicide mobility in soil

We have listed here the most important concepts to understand the methodologies involving the mobility of herbicides in the soil, according to the literature and international guidelines established for these studies (EPA 1998, 2008; OECD 2000, 2004).

> *Transport:* Process that involves the movement of the herbicide in different environmental compartments, such as drift, volatilization, runoff, and leaching. It can occur in different directions and intensities, from which the product can be taken to other locations. As in

the case of herbicides, leaching is the main process of transporting
these molecules from the soil to groundwater.

Mobility: The mobility of an herbicide consists in its ability to move
within an environment (environmental compartment), and therefore it is important to consider it as a determinant for the transport
of the herbicide and its degradation products (FAO 1989). This process is closely linked to the physical-chemical characteristics of the
herbicide and the soil and provides information on the distribution
of the product in the soil over short distances.

Leaching: Process by which the herbicide moves vertically along with
the water flow, either in an unformed soil profile in field conditions
(lysimeter) or glass/PVC column filled with deformed (sieved) soil
in a laboratory or greenhouse.

Leach distance: Deeper segment (layer) in the soil profile or column in which
the herbicide was able to reach, with at least 0.5% of the total applied at
the beginning of the study (equivalent to penetration into the soil).

Leachate: A water phase that percolates with the herbicide in the soil
profile or in the extension of a glass column and is collected at the
base of the structure with a container at different times.

Mass balance: Sum of all the results of the amounts of the herbicide
found in the soil layers and in the leachate (percentage of radioactivity in relation to the total applied) obtained, in order to verify the
consistency of the results and the conservation of mass in relation to
that applied initially in the studied system.

Artificial rain: Solution of 0.01 M $CaCl_2$ in distilled or deionized water
that is applied by a simulator (peristaltic pump) on top of the soil
column throughout the experimental period.

Lysimeter: Instrument used in the measurement of plant evapotranspiration. For mobility studies, this instrument is used to verify the leaching of pesticides in the soil profile in representative columns of the
soil, inserted in a given cultivation system. The instrument has channels in which the leachate herbicide is collected along the soil profile.

Glass or PVC column: Used in column leaching studies, they represent
30 cm of soil profile (deformed soil samples). They allow the installation of a simulated rain system, application of radiolabeled herbicides (without adhering to the column walls) and measurement of
the percentage of herbicide leached in sections in the soil.

Soil TLC: Thin layer chromatography of soil. The method consists of
glass plates, which have a thin layer of soil, proposed in the methodology of Helling and Turner since 1968. This chromatographic
technique (separation of mixtures) depends on the affinity of the
herbicide with the components of the stationary phase (soil) and
mobile phase (water), from which the mobility of herbicides in the
soil is determined over short distances, by the value of *Rf.*

Elution: Processor that promotes the dragging of the applied substances on TLC plates by means of a solvent. In mobility studies the solvent used is water, which represents the process of displacing the herbicide in the soil (soil plates) over short distances, promoted by water.

Autoradiography: Qualitative analysis performed to identify the behavior of pesticides in plants and soil. In soil TLC mobility studies, images are generated indicating the path the herbicide took after plaque elution.

Chromatogram (soil TLC): Graph indicating the intensity (peaks) of the herbicide radioactivity in relation to the distance covered by the product along the soil TLC plate.

Rf: Retention factor, which is the longest distance covered by the herbicide in a soil TLC plate divided by the distance covered by a solvent front (deionized water) at a distance of 10 cm.

3.2   Mobility of ^{14}C-herbicides in the soil by the thin-layer chromatography method (soil TLC)

According to the EPA (1998), soil TLC is an appropriate qualitative tool to obtain an estimate of the mobility potential of a given herbicide over short distances. This tool was introduced to study the mobility of organic chemicals in the soil in the 1960s by Helling and Turner (1968) as an alternative to column analysis and sorption isotherms in the soil. In this method, the conventional stationary phase (gels and silica, oxides, etc.) is replaced by a thin layer of soil.

Soil TLC studies offer many advantages, since they are reproducible and allow the application of statistical analysis. It also can allow the distinction between the components of mass transfer and diffusion and provide information about the chemical transformations of the product in the soil (EPA 1998). On the other hand, by estimating the mobility of the product in soil plates horizontally, the method does not allow one to determine mobility in depth, so that methods such as leaching in soil columns and lysimeters in the field may be more suitable for this purpose.

Soil TLC can be used to determine the mobility of herbicides from moderate to very soluble in water. It is noteworthy that products with water solubility below 0.5 ppm do not need to be tested, as these generally do not have mobility in the soil (EPA 1998). The mobility of volatile products can also be assessed by this method.

3.2.1   Preparation of soil TLC plates

Before the mobility study, it is important to obtain information about the physical-chemical properties of the soil to be studied, such as texture,

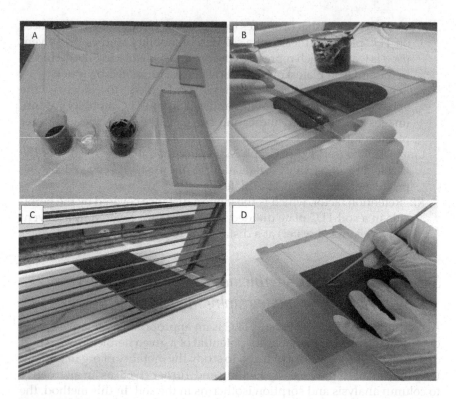

Figure 3.1 Methodological sequence for making soil TLC plates. Soil paste to apply on the glass TLC plate (A); making the thin layer of soil on the glass TLC plate (B); prepared plate, drying at room temperature for later use in herbicide mobility study in TLC soil plates (C); marking of the horizontal line that delimits the eluent (D). (From the collection of the Ecotoxicology Laboratory, Center for Nuclear Energy in Agriculture [CENA/USP], Piracicaba, Brazil.)

sample location, sampled horizon, particle size, percentage of organic matter, pH, humidity, capacity cation exchange, among others.

The soil must then be dried at room temperature, ground, and sieved through a 250-μm granulometric sieve to reduce the size of the aggregates. The sieved soil should be mixed with water until a moderately fluid paste is created (approximately 0.75 mL of water should be used for each gram of soil) (Figure 3.1A). Subsequently, this soil paste must be spread evenly on a TLC glass plate with the aid of a spreader or a glass rod (Figure 3.1B). The thickness of the soil layer must be thin and vary between 0.5 to 0.75 mm.

The plates must be placed to dry at room temperature for a period of approximately 24 h. It is recommended that triplicates be used for each type of soil or treatment evaluated and that all procedures be performed at 23 ± 5°C (EPA 1998).

3.2.2   Application and elution of ^{14}C-herbicide

At a height of approximately 11.5 cm from the base of the plate, a horizontal line must be drawn to interrupt the movement (running) of the eluent (Figure 3.1D). The ^{14}C-herbicide must be applied with the aid of a microsyringe at a height of 1.5 cm above the base (0.5–5 mg of the active ingredient containing 0.01–0.03 mCi of ^{14}C); and the product will elute in total 10 cm from the plate, as shown in Figure 3.2A and B. If necessary, the same concentration of the analytical standard of the herbicide can be applied next to the sample as a reference.

Then the plate must be placed vertically in a closed chromatographic chamber (glass vat), containing the mobile phase at a height of 0.5 cm (EPA 1998) (Figure 3.2C). The mobile phase must be deionized water or another solution in which it is desired to determine the mobility of the product by the soil; however, water is the most suitable solvent to simulate

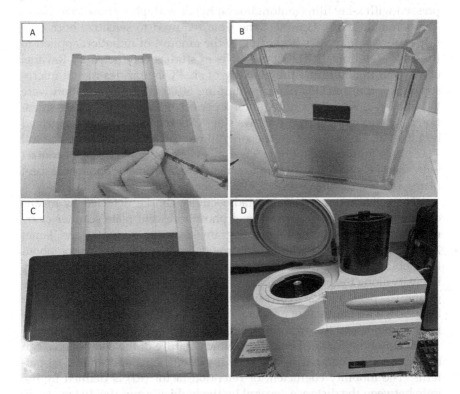

Figure 3.2 Delimitation of the application lines of ^{14}C-herbicide and solvent top (A); application of the working solution with the aid of a microsyringe 1.5 cm from the base of the soil TLC plate (B); TLC soil plates placed to elute in a vertical chromatographic chamber (C) and the zones containing radioactivity are detected by reading the TLC soil plates on phosphorusplate scanners (Radio Scanners) (D).

the movement of rain in the soil microstructures in the field (Ravanel et al. 1999). If the herbicide is volatile, it is essential that a glass plate be placed over the soil TLC plate to prevent volatilization.

At the end of the elution, when the mobile phase reaches the limit line at the top of the soil plate, it must be placed to dry at a temperature of 25°C for 24 h. However, if there are volatile components, the plate must be immediately cooled (Helling 1971). After applying and drying the plate, x-ray images are taken so that it is possible to infer about the mobility of the herbicides (Figure 3.2C and D).

3.2.3   *Autoradiography or radiocromatogram of soil TLC and retention factor of* ^{14}*C-herbicide*

To determine the path taken by the herbicide, the soil TLC plates must be pressed with x-ray films (autoradiography) or with phosphorescent plates (radiocromatogram). However, the time required to sensitize both contact materials may vary according to the amount of radiation applied to the plates (Helling 1971). For example, in studies conducted by Ravanel et al. 1999) to determine the mobility of six ^{14}C-herbicides in soil microstructures, the soil TLC plates containing 50,000 dpm of radiation were left in contact with the x-ray film for a period of 3 days. Shorter sensitization time was applied in the analysis of ^{14}C-pesticide mobility performed by Liu et al. (2018), in which the x-ray films were in contact with the soil TLC plates for a period of only 2 h. However, a 24-hour period of contact with soil TLC plates is sufficient to sensitize x-ray films or phosphorescent plates, in most mobility studies with ^{14}C-herbicides.

Then, the zones containing radioactivity can be detected by means of autoradiographies generated by a phosphorescent plate scanner or by means of scans with radiocromatogram scanners. (Cornejo and Jamet 2000). These reading methods can provide qualitative information about the mobility pattern, as well as chromatograms with peaks of radioactivity intensity along the path taken by the herbicide. In the generated chromatograms, the peak intensity corresponds to the radioactivity intensity of the herbicide, with the areas of greater radioactivity on the plate showing more intense colors, and the opposite occurs for the areas of less radioactivity.

The mobility pattern and the chemical nature of the components can be identified by comparing the mobility coefficients and analytical patterns. The mobility coefficient or retention factor (Rf) is defined by the ratio between the distance covered by the herbicide and the distance covered by the solvent, as shown in Equation 4.1 and in Figure 3.3.

$$Rf = \frac{\textit{Distance from baseline traveled by herbicide (a)}}{\textit{Distance from the baseline traveled by the solvent (b)}} \qquad \text{Eq. (4.1)}$$

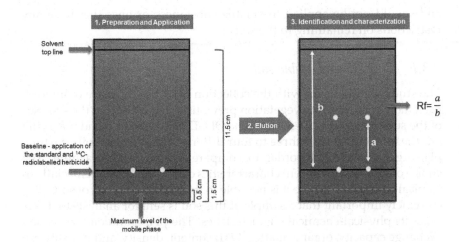

Figure 3.3 Schematic illustrating the dimensions and steps in the application of ^{14}C-herbicide on TLC soil plates, where "a" corresponds to distance, from the baseline covered by the herbicide and "b" corresponds to distance from the baseline traveled by the solvent.

For a more detailed analysis, it is recommended that the plates be photographed in order to improve the visualization of the mobility pattern. Later, the studied herbicide can also be extracted from the accumulated point and submitted to a recovery analysis.

3.3 Leaching of ^{14}C-herbicides in soil columns

The Organisation for Economic Co-operation and Development (OECD) is a world reference in relation to studies using herbicides, and in the case of leaching studies in soil columns, there is a very well-established protocol that must be carefully observed before starting any study: the OECD 312 "Leaching in Soil Columns" (OECD 2004). However, there are other materials that can help and that bring important complementary information to the study, such as the Environmental Protection Agency (EPA) (2008) and the literature reviews of Katagi (2013) and Mendes et al. (2017). As each study has a unique approach, it is essential that all these materials are consulted.

Katagi (2013) points out that the study of column leaching can assist in the determination of several key parameters for the mobility of herbicides, such as the effects on leaching due to soil disturbances, sorption imbalance, preferential transport, and biodegradation of molecules.

The following description of the procedures was based on the guidelines of the OECD (2004). According to the guidelines, this method is applicable for radiolabeled substances, with sufficient accuracy and sensitivity, and should not be applied to volatile substances or ones that have any restrictions on remaining in the soil.

3.3.1 Collection of arable soil

The study should begin with the collection of soil in the place of interest, preferably removing the vegetation previously and collecting the samples of the superficial layer (0–20 cm). The OECD (2004) recommends carrying out the test with at least three to four different agricultural soils (texture, pH, organic matter), in order to compare the product's leaching. The choice of soils with different characteristics can be important, especially in tropical conditions, where it is possible to find different types of soils. It is extremely important that a sample of this soil is sent for analysis to determine its physical-chemical characteristics. The pH, granulometry, cation exchange capacity, organic matter (OM) content, density, and porosity are just some of the parameters that can support the leaching of herbicides in the soil column (Oliveira and Brighenti 2011). The soil samples must be dried for at least 1 week at room temperature before being sieved (<2 mm) and stored in a freezer (<4°C) until the time of the study.

The study should be carried out in an air-conditioned place without the presence of light, and temperature between 18 and 25°C (OECD 2004). However, these conditions can be modified to better represent the environmental conditions of the field and can be shaped according to the interest of the study.

3.3.2 Preparation of soil columns

The stage of preparing the columns is very important in the leaching studies, as the correct preparation certainly adds quality to the results. The sieved and dried soil is evenly distributed and packaged along the column profile so that there are no preferential paths for water flow in the soil profile.

The recommendation of the OECD (2004) points to the use of radiolabeled molecules for the study, usually using glass columns, as will be described in the next topics. It is also possible to use stainless steel, Teflon or polyvinyl chloride (PVC) columns, which are suitable for this type of study (Katagi 2013). In order to choose the material to be used in the column, it is ideal to analyze the possible sorption of the herbicide of interest with the most common materials. Koskinen et al. (1999) reported only a slight sorption of some ^{14}C-herbicides in a PVC column, such as atrazine, alachlor, and nicosulfuron. Specifically, in the aforementioned cases and in general, glass columns are more suitable for the vast majority of studies, as they do not interfere in the leaching of most herbicides.

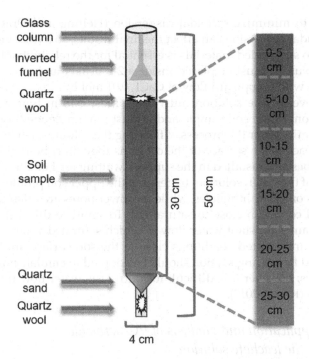

Figure 3.4 Schematic illustrating the dimensions, components, and sections of the glass columns for leaching studies of ^{14}C-herbicides.

The columns used for leaching studies are, in most cases, 0.5 m long and about 4 cm in diameter (OECD 2004). As already determined by the OECD for studies using radiomolecules, two repetitions are sufficient for validation of the studies, in order to generate the least amount of radioactive waste possible.

After preparing the columns, the first step is to fix them in a suitable support so that they are vertically aligned and firmly adhered. The bottom opening of the column should be filled with quartz wool, and the tapered part just above should be filled with washed quartz sand and kiln dried at 100°C. After the sand, the soil can be placed, filling the body of the column up to 0.3 m in length, varying according to the objective of the study (Figure 3.4). The mass of the soil samples inside the columns must be measured so there will be a control over the reproducibility of the process.

Before applying the herbicide, it is necessary to moisten the desolate column by capillarity. This procedure is necessary so that there is no preferential flow of leaching of the herbicide during the simulation of rain during the study evaluations. Elution should take place with distilled or deionized water and, preferably, with electrolytes such as 5 to 10 mM $CaCl_2$

or $CaSO_4$ to minimize colloidal dispersion (Helling and Dragun 1981). This colloidal dispersion can affect the final results, so a translucent sample with no suspended materials is essential for the reliability of the studies. The columns must be placed inside a 2-L beaker and must be slowly moistened with an upward flow of $CaCl_2$ 0.01 mol L^{-1}.

The level of the solution must be at most 0.10 m higher than the wetting front of the soil sample and it must remain flooded for approximately 30 min. After the process of flooding the columns, when the $CaCl_2$ solution reaches the surface of the columns they must be removed from the test tubes and installed in the support, waiting for 1 or 2 h for the total drainage of the $CaCl_2$ solution. In general, the upper soil profile is unsaturated and contains air bubbles, while the lower pores are often filled with water (soil condition close to saturation). To simulate this last situation, just maintain a constant water flow in which saturated water conditions persist. If unsaturated conditions close to the soil surface are tested, an interrupted flow using suction should be adopted to emulate normal field water flows; however it is difficult to control the water flow under these conditions (Katagi 2013).

3.3.3 Application and analysis of ^{14}C-herbicide in the leachate solution

After draining, there are two main methods for applying the herbicide: application on dry or wet soil. These approaches are intended to simulate a situation in which there was rain close to the application period (wet soil) or during a period with little rainfall (dry soil). If it is on dry soil, the ideal is to remove approximately 100 g of soil from the surface of the column before wetting and apply the herbicide in this portion, homogenize, and return the soil to the column only after the flooding process is over (OECD 2004).

The application process in humid soil is simpler and is used the most, since the herbicide can be applied directly to the column surface, also after the process of capillary wetting with $CaCl_2$, without the need for the previous removal of a portion of the soil.

Herbicides must be applied in the highest recommended dose for the culture in which they were registered, so the amount of radioactive product applied to the column must be sufficient to allow at least 0.5% of the dose applied in any segment at the end of the study, with the herbicide showing 95% radiochemical purity (OECD 2004). Generally, about 1,000,000 dpm per column (16666.66 Bq) is used, making it possible to analyze layers easily.

The application solution, made with the solvent that provides greater solubility (for example, with the use of saflufenacil, acetone is recommended), must consist of 200 to 250 μL in total (herbicide + solvent), containing the radiolabeled herbicide, respecting the field dose, which must be added with the analytical standard of the herbicide. This process helps to reduce the amount of the radioactivity used, avoiding possible contaminations. The solution must be applied with the aid of an automatic pipette directly on the moist soil on the surface of each column, or on the sample of 100 g of dry soil that was previously separated during the process of accommodation of the column, with subsequent homogenization of the soil inside the column. As reported by Isensee and Sadeghi (1992), there were no statistical differences in the leaching of two dyes that were applied in the center or on the edge of a 0.10 m diameter soil column, but it is preferable that during application the herbicide be distributed as evenly as possible on the column surface.

After the application of the herbicide, the surface of the soil sample must be covered with a glass wool disc, in order to dissipate the simulated raindrops, by fitting an inverted funnel. In this funnel, a plastic tube must be connected, through which the 0.01 mol L^{-1} $CaCl_2$ solution (simulated rain) will be conducted. For the simulated rain process, it is recommended to use a peristaltic pump above the soil columns. A flow of approximately 8 mL h^{-1} must be simulated for 48 h, resulting in a simulation of approximately 200 mm rain, based on a column of 4 cm in diameter (OEDC 2004). Despite the high amount of simulated rain, this condition is used frequently; however it must be adapted according to the objective of the study.

Flasks of the Erlenmeyer or Schott type must be positioned just below the columns, becoming the container for the collection of the leached solution from each of the columns (Figure 3.5). It is recommended that collections be carried out at an interval of no more than 12 h. Collections should take place in triplicates, with 10 mL aliquots each in vial-type flasks. Due to the high sampled volume of an aqueous solution (10 mL), it is essential to use the scintillating solution of the type Insta-gel, which supports this type of sampling, for further analysis in the liquid scintillation spectrometer (LSS) for 15 min.

3.3.4   Analysis of ^{14}C-herbicide in soil layers

At the end of 48 h, the simulated rain must be discontinued and the soil layer samples removed. The glass columns must be carefully removed from the support, and it is necessary to inject an air flow at the bottom end of the column to force the soil out through the top of the column. The samples are normally sectioned in six layers of 5 cm each. Each layer of soil is packed in trays for drying at room temperature. When it is not possible to carry out this process, the column must be frozen after the water is completely drained so that the conditions of the sample are preserved in full and do not break the glass column.

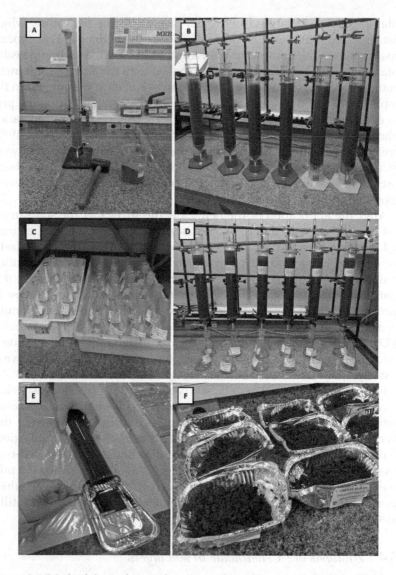

Figure 3.5 Methodological procedure in studies of leaching of 14C-herbicides in glass columns. Soil accommodation inside the glass column without bubbles and preferential paths (A), saturation of the columns with $CaCl_2$ solution (0.01M) in the preparation of the column to receive the ^{14}C-herbicide (B), Erlenmeyer flasks for leachate collection during the study period (C), columns after the herbicide application, on supports, with simulated rain application and leachate collection (D), sectioning the soil profile of the columns after 48 h of simulated rain (E), and drying of the soils of the column sections for later quantification of radioactivity (F). (From the collection of the Ecotoxicology Laboratory, Center for Nuclear Energy in Agriculture [CENA/USP], Piracicaba, Brazil.)

The soil of each layer must be ground and homogenized, and the total mass must be measured. To quantify the radioactivity from the ^{14}C-herbicide, three 0.2-g aliquots must be separated in porcelain containers to be oxidized in a biological oxidizer for 3 min, to then be quantified by LSS for 5 min. Through the combustion of the soil, the radiomolecules contained in the material are transformed into $^{14}CO_2$, being conducted by the system by a continuous flow of O_2 and N_2 to a flask containing scintillating solution, which traps the $^{14}CO_2$ to be quantified later.

If there is interest in checking whether the molecule has degraded during the leaching study, it is possible to pass the soil through an extraction process with specific solvent for each herbicide in order to determine the parent molecule and its possible metabolites. The TLC separation method is recommended for this case (OECD 2004).

It is important to note that studies on leaching in soil columns do not fully quantify the potential for leaching of the herbicide in the field. However, the results obtained can assist in decision making regarding chemical management of weeds, leaching in the soil, and possible ecotoxicology to aquatic and terrestrial organisms to serve as a basis for requesting further studies under field conditions.

The results obtained are expressed as a percentage of the radioactivity found in the leachate in each of the spine segments in relation to the initial applied radioactivity. According to the OECD (2004), the mass balance in these studies (sum of the percentages of ^{14}C-herbicides found in soil layers and leachate) can vary between 90% and 110% of the total herbicide applied at the top of the column.

3.4 Leaching of ^{14}C-herbicides in lysimeters

3.4.1 Site preparation and study materials

Lysimeters are large perforated tanks inserted in the field and filled with soil to measure evapotranspiration and the percolation of organic compounds in the soil, being a highly recommended tool for the study of herbicide leaching (Figure 3.6). Through channels along the equipment it is possible to collect the leached solution at different depths of the soil. The different collection points form a diagonal depth gradient, and these points are in a corridor that allows access for collection (Führ et al. 1991). It is important to remember that the physical and chemical characteristics of the soil inside the lysimeter are very similar to the conditions of the soil around the equipment.

With the study being carried out in the field, the first step is to determine an adequate area, avoiding the presence of nearby springs and an area with a history of herbicide application. Then the study of the characteristics of the soil, such as humidity and temperature, which can

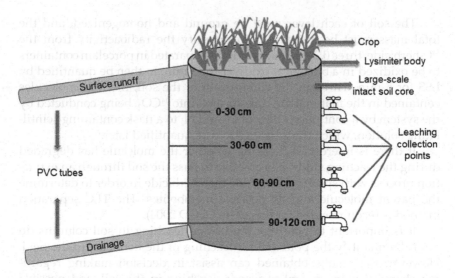

Figure 3.6 Schematic illustrating the size and composition of a lysimeter installed in the field for leaching studies of ^{14}C-herbicides.

considerably influence the analyses, should be done. Then the lysimeter must be accommodated in the desired collection site and filled with the local soil, as previously described.

Other tools, such as tensiometers, can also be used in conjunction with lysimeters to help control variables. The tensiometers can be used for direct measurement of matrix suction (Diene 2004), and consist of a porous element, a cylindrical body, and a pressure sensor. It is also necessary to know the water depth required for each culture, according to the interest of the study. There are also lysimeters equipped with load cell weighing mechanisms. This equipment makes it possible to obtain measurements on an hourly scale with great precision (Carvalho et al. 2007).

3.4.2   *Application and analysis of ^{14}C-herbicide*

Once the equipment is installed in the field, the application of the radio-labeled herbicide must be performed by spraying, avoiding product loss through drift, which can occur mainly through volatilization and displacement to other areas by means of the wind, increasing the cost of the study and causing possible contamination in the environment. Thus, climatic conditions are ideally stable during application, without wind, and with ambient temperature and low humidity, aiming to reach the soil in the most uniform way possible (Kördel et al. 1991). The collection time depends on the physical-chemical characteristics of the herbicide and the time spent in the soil (persistence).

For example, Dörfler et al. (2006) applied the herbicide ^{14}C-isoproturon in a lysimeter, simulating intense rain conditions. The collections of the leached solution were performed weekly due to the moderate persistence of the herbicide, which can range from 60 to 300 days (Schroll et al. 2006). After collection on the chosen days, the solution must be immediately analyzed; otherwise, it must be stored in a refrigerator at 4 ± 2°C. In the case of an herbicide such as saflufenacil, with a shorter half-life of 20 days (PPDB 2020), collection intervals should be reduced.

In general, lysimeters can be used for leaching studies up to a depth of 3 m, but it is possible to use even larger systems if one is interested. Depending on the simulation, different densities of the soil contained in the lysimeter are used. As a parameter, 115 kg for each 0.10 m depth should be used. Regarding the collection of leachate material, taps are used in lysimeters, being positioned on a descending ladder according to the desired analysis depth.

The mass balance of this type of study can vary between 90% and 110% in relation to the total radioactivity applied (OECD 2004). It is important to note that the entire experimental design must be done with the previous analysis of the physical-chemical characteristics of the soil and also of the product properties. Information about biodegradation and leaching studies on soil columns of the herbicide of interest in a similar soil can provide fundamental information for the study.

3.5   Purpose and choice of methods used to characterize the mobility of ^{14}C-herbicides in the soil

The studies described above have different methodologies and dimensions, with an increasing order of complexity, which must be aligned with the objectives of the studies to be carried out. The soil TLC method allows a direct classification of mobility by the classes developed by Helling and Turner (1968). A great advantage is the possibility of differentiating the components of mass transfer and diffusion, in which the data can be subject to statistical analysis. Another advantage is due to the small amount of soil to be used, enabling the comparison between different types and depths of soil, in which it makes it possible to evaluate the herbicide-soil interaction.

On the other hand, the method of soil TLC is not consistent for volatile herbicides; due to the conditions under which the study will be conducted, the mobility of the product will not be representative. It is worth mentioning that the mobility estimated in the soil TLC does not present data of mobility in depth of the soil, unlike the methods of leaching in glass columns or lysimeters in the field, which may be more appropriate in these cases.

Column herbicide leaching studies enable analysis under controlled laboratory conditions, predicting the leaching potential before conducting studies in real field conditions. This inaccuracy in relation to the conditions in the field is due to the fact that the soil is unstructured and uniform (sieved, dry and without vegetable residues), packed in glass columns, thus representing only a demonstrative unit. On the other hand, this type of study does not require a large amount of radiolabeled material to be carried out, avoiding excess of solid waste, and can be carried out at low costs.

Studies carried out on lysimeters with the use of radiolabeled molecules presented greater precision and sensitivity of the analyses. The advantages of the mobility study by lysimeter are in the fidelity of the results obtained, as the equipment is installed in the field, making the climatic conditions, cultivation, and cultural residue at the time of the study, structural characteristics of the soil, and degradation of the herbicide, as real as possible. However, it is necessary to use a large area and a large amount of radiolabeled material, which increases the costs and the amount of solid waste in the study. Thus, the method performed under field conditions is more complex, and to avoid the unnecessary use of radiolabeled material and the expenditure in the implementation of the study in lysimeters, the physic-chemical properties of molecules should be known, as a leaching potential, Groundwater Ubiquity Score (GUS) index, sorption–desorption coefficients, vapor pressure that estimates volatilization, and half-life of degradation in the soil.

For herbicides with a high leaching potential, Laabs et al. (2002) described the presence of a series of molecules in the soil profile and in the leachate, collected in lysimeters with 95 cm depth, a length that would be practically impossible to perform in soil columns in the laboratory. These molecules, such as metolachlor, alachlor, atrazine, and simazine, are more suitable for lysimeters due to their high leaching potential (Laabs et al. 2002), allowing a better quantification in the deeper layers of the soil profile.

Each study, as described, has specific characteristics that demand factors and resources intrinsic and extrinsic to the chosen method, which the researchers must consider before the study is carried out. Figure 3.7 illustrates a comparative scheme between the three methods explained above, and dimensioning in their respective scales.

3.6   How to report the results obtained in mobility studies?

The results for mobility studies using radiolabeled herbicides should be reported following the guidelines established. In soil TLC, according to EPA (1998), the results should contain the amount of the chemical applied

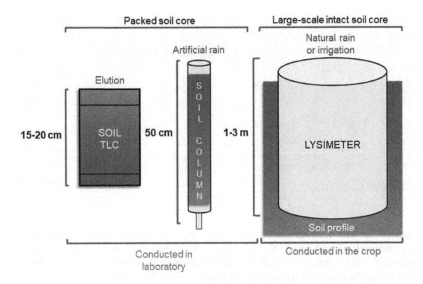

Figure 3.7 Illustrative scheme comparing dimensions and application conditions for ¹⁴C-herbicide mobility methodologies. (Adapted from Katagi 2013.)

and the amount recovered from the plates. It should also contain the average "frontal" Rf value with standard deviation for each soil tested and perform an autoradiography or diagram of the soil TLC plate that shows all the movement of the herbicide (from 1.5 to 11.5 cm).

Reports of studies on TLC soil plates can be performed by combining autoradiography and chromatogram, as shown in Figure 3.8, generated at the Ecotoxicology Laboratory, Center for Nuclear Energy in Agriculture, CENA, University of São Paulo, USP, Brazil. This study was published by Mendes et al. (2019a), in which they assessed the mobility of ¹⁴C-pendimemethalin and ¹⁴C-diclosuram in soil plus 0.1% of biochar, obtaining the values of $R_f = 0.2023 \pm 0.0032$ (Figure 3.8A) and $R_f = 0.9921 \pm 0.0022$ (Figure 3.8B), respectively. In this research, the authors indicated only the movement of the products on the plates; the chromatogram images generated were added to the collection of the Ecotoxicology Laboratory, CENA/USP.

In the case of leaching studies carried out in columns using radiolabeled herbicides, the data report must follow the guidelines suggested by OECD (2004), which can also be applied to field studies by lysimeter for standardization and later comparisons between studies. In the guidelines, the results must be reported in a table which shows the concentration in percentage of the applied dose for each layer of the soil and the leached solution, appropriate mass balance, volumes of the leached solution, leaching distance and, if appropriate, relative mobility factors, figures of

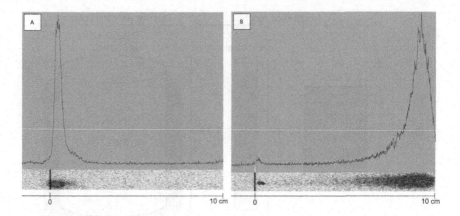

Figure *3.8* Chromatogram indicating the intensity and mobility of ^{14}C-pendimethalin (A) and ^{14}C-diclosuram (B) in soil TLC (treatment correspond-ing to soil + 0.1% natural biochar). The distances represent the extension from of point of application to elution line in plates of soil TLC. (Adapted from Mendes et al. 2019a.)

the percentages of herbicides found in soil layers in relation to soil depth, and discussion and interpretation of data.

Following the same fundamentals as the OECD guidelines (2004), Mendes et al. (2018) developed a scheme relating the soil depth in a glass column to the initial percentage applied to the herbicide ^{14}C-aminocyclopyrachlor (Figure 3.9A). Guimarães et al. (2019) also reported the leaching data in columns using graphs; however, it indicated in bar graphs the distribution of metribuzin in different soils (Figure 3.9B). Figures that indicate leaching along the soil profile in a didactic way are strongly recommended, as they indicate to the reader a more representa-tive idea of what occurred throughout the study.

For field studies with lysimeters, the necessary information is similar to column leaching studies; however, data about the field, such as type of culture, cultural practices in implementing the system, history of applica-tion of the area, geographical location, environmental conditions of tem-perature and precipitation, irrigation used, and data on the degradation of the herbicide in the field, must be included when reporting the results (OECD, 2000). Table 3.1 shows part of the results of the studies carried out by Grundmann et al. (2008), who evaluated the leaching of ^{14}C-glyphosate in five agricultural soils, with soybeans grown at the time of the study. In this study, the table was chosen as a way to report the leaching data in a lysimeter, which in this specific case presented additional information, such as volatility and mineralization of the herbicide, which can assist in the discussion of the data.

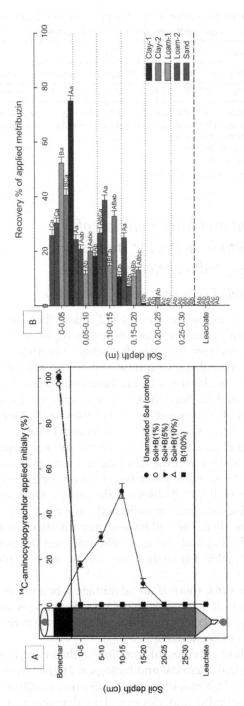

Figure 3.9 Evaluation of the herbicide ¹⁴C-aminocyclopyrachlor in soil with different levels of bonechar (A) and the herbicide ¹⁴C-metribuzin in different types of soil (B), indicating the percentage of herbicides along the soil profile in glass columns. (Adapted from Mendes et al. 2018 (A) and Guimarães et al. 2019 (B).)

Table 3.1 Mass balance 48 days after the first application of ^{14}C-glyphosate to the lysimeters (percentage of applied radioactivity)

	Soil			Plants				Mass
	$^{14}CO_2$[a]	Volatility[a]	Residues	$^{14}CO_2$[a]	Volatility[a]	Residues	Leachate	Balance
Lysimeter	38.55[b]	<0.001	53.20	2.10[c]	<0.001	5.71	<0.001	99.56

[a] Mineralization and volatilization data for soil and plant surfaces were obtained from two chambers in the soil and two chambers in the plants, and the data provided are therefore the average of two single measurements.
[b] Average value of 38.86% and 38.23%.
[c] Average value of 2.42% and 1.78%.
Source: Adapted from Grundmann et al. (2008).

3.7 Presentation and interpretation of literature data in herbicide transport studies using radiolabeled molecules

3.7.1 Mobility in soil TLC plates in the laboratory

The mobility studies of herbicides by soil TLC allow the calculation of a basic parameter to analyze the results, known as Rf, which varies from 0 to 1. This value results from the ratio that relates the distance from the baseline traveled by the substance (phase stationary) over the distance from the baseline traveled by the solvent, the mobile phase (Sherma and Fried 2003) (as indicated in Section 3.3). Each herbicide molecule will have traveled a different distance at the end of the chromatographic run, which will correspond to a specific Rf value related to each type of soil. According to Ravanel et al. (1999) the greater the soil sorption forces for the applied substances, the lower the Rf value and with that, a theoretical correlation between the Rf values and the sorption coefficients (K_d) can be established. Bukun et al. (2010) observed the same correlation between K_d and Rf for the herbicides aminopyralid and clopyralid, indicating that sorption and an herbicide in the soil have a relevant effect on the mobility of an herbicide. Helling and Turner (1968) indicated the classification of herbicide mobility (Table 3.2) so that they can be used to interpret the study results.

Added to these results, there is an advantage when using a radiolabeled herbicide, because it is possible to obtain x-ray images (autoradiography) by means of light intensity, the concentration of the herbicide in the soil TLC, as described in Section 3.3.

The soil TLC method provides information on the interaction between soils of different physical-chemical and biological characteristics, as well as different herbicides. In a mobility study, Bukun et al. (2010) evaluated the herbicides aminopyralid and clopyralid in five types of soil, with

Table 3.2 Values of retention factors (Rf) and mobility potential
of herbicides in soil TLC

Rf values	Class	Mobility potential
0–0.09	1	Not mobile
0.10–0.34	2	Low mobility
0.35–0.64	3	Moderate mobility
0.65–0.89	4	Mobile
0.90–1.00	5	High mobility

Source: Adapted from Helling and Turner (1968).

pH variation from 5.4 to 8.0 and OM from 1.2% to 7.9%, and a significant correlation was observed between mobility and OM, as it increased the sorption of both herbicides to the soil. On the other hand, in the study by Guo et al. (2003) on the mobility of fomesafen, it was verified that the pH of the studied soils and the physical-chemical properties of the substance, more specifically the dissociation constant (pKa = 2.7), played an important role in mobility, because its predominant anionic form decreased sorption of the herbicide to the soil.

Jacobsen et al. (2001) evaluated the mobility of atrazine by soil TLC. The study evaluated different layers of soil from an agricultural area with a history of maize cultivation, and the results indicated that in the superficial layers in which they had a greater amount of OM, the Rf value was 0.10; in contrast, in the subsurface layers it was 0.81. With that, it was concluded that the higher OM content decreased the leaching potential of atrazine and the risk of contamination of groundwater, due to the increase in the sorption of the herbicide to the soil components.

The advantage of studying herbicide mobility by soil TLC is the reproducibility of results and the ease of the technique, as seen in a study by Strebe and Talbert (2001), who evaluated the mobility of flumetsulam in 13 soils at 2 depths (0–15 and 30–46 cm) from the southern United States. In this study, they found that the herbicide showed high mobility in all soils, and the determining factor was the pKa of the herbicide, which behaves like a weak acid, predominating its anionic form and decreasing sorption in the soil.

Outside the agronomic, Flessner et al. (2015) reported another study that also used the soil TLC method scope. The objective was to assess the mobility of methiozolin and isoxaben in sandy soils, where golf course lawns are installed. The results indicated that the Rf values of the mobility of methiozolin and isoxaben were 0.46 and 0.10, respectively, due to the hydrophobic and low S_w chemical characteristics of the herbicides.

Deng et al. (2017) used the soil TLC method to assess the potential of atrazine retention in the soil after application of biochar produced with cassava residues. Artificial rain was applied, and soils with different proportions of biochar (0, 0.1, 0.5, 1, 3, and 5%) were used as a stationary phase. The results indicated that the addition of the biochar significantly increased the sorption capacity of atrazine in the soil according to the rates of application of the biochar. However, the authors also observed that the magnitude of the herbicide retention depends on the pH of the solution, the temperature of the environment, and the persistence in the soil.

Also using biochar, Jones et al. (2011) evaluated the behavior of simazine applied to soil with biochar, based on soil TLC. Soils with high and low fertility, artificial rain with distilled water, and three different biochar arrangements were used where the herbicide was applied. The results obtained indicated that, in general, biochar suppressed biodegradation and reduced leaching of simazine, due to the fast and high sorption capacity of the material. Still using biochar, Mendes et al. (2019b) evaluated the adsorbent effect of a doll (bone coal [BC]) for quinclorac, metribuzin, and hexazinone in a tropical soil. TLC determined the mobility of the molecules in soil in four different treatments: soil not altered, soil containing 5% BC, soil with BC band (1 cm) in the middle of the plate, and soil with BC band at the top. Figure 3.10 shows images of some treatments from the previous study and their respective autoradiographies, where it is possible to visualize the distance covered by the herbicides as well as their concentration by light intensity. The results obtained showed that the BC has a high sorption capacity and is excellent for reducing the mobility of these herbicides in the soil regardless of the form of application. As demonstrated in this study, the TLC method in soil allows different treatment arrangements for the stationary phase and variations in the composition of the mobile phase, allowing a better understanding of the process of mobility of the herbicides in the soil.

Additional information on the method and applications of TLC on the ground can be found in the OPPTS guidelines 835.1210 established by the EPA (1998) and review articles by Sherma and Rabel (2018) and Helling and Turner (1968).

3.7.2   Leaching of herbicides in columns in the laboratory

The study of column leaching makes it possible, in addition to the use of different types of soil, to modify the surface fraction at the top of the column, simulating environmental conditions and agricultural practices, such as the use of organic materials, agro-industrial residues, and aging of the herbicide in the soil, for example. In a study by Si et al. (2005), the leaching of ethametsulfuron-methyl was evaluated in three different soils, using aged soil surface (30 days) with ethametsulfuron-methyl and the other not aged. The results indicated that in addition to the effect of

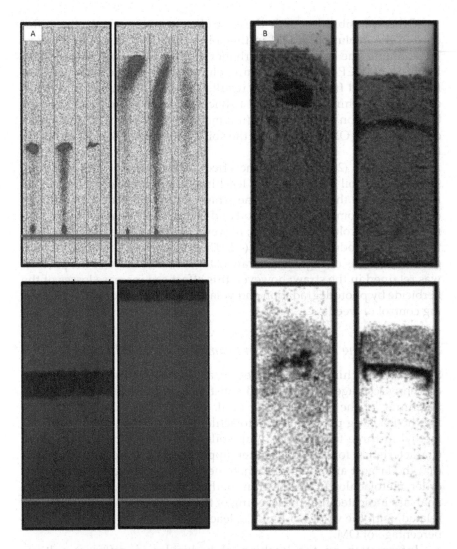

Figure 3.10 Presentation of the application of organic material in TLC soil plates. Mobility of ^{14}C-herbicides quinclorac (left), metribuzin (middle) and hexazinone (right) in plates containing soil + bonechar. (A). Example of mobility of ^{14}C-simazine in plates containing soil + biochar (B). (Adapted from Mendes et al. 2019b and Jones et al. 2011.)

OM, clay content, and soil pH, aging significantly increased the percentage of herbicide retained in the soil. This occurred due to the increase in herbicide retention on the colloid surface and occupation of the sorption sites, concluding that for most herbicides, sorption can limit the leaching process.

Other possibilities with these studies are to evaluate the influence of the evolution of OM in the soil on the leaching of herbicides.
For example, the study by Haberhauer et al. (2002), who evaluated the
leaching of MCPA in soil columns eluted with different fractions of
humic acid and fulvic acid. The results showed that humic acid accel-
erated the leaching process and fulvic acid retained more of the herbicide in relation to the control (unamended soil), concluding that the
composition of OM dissolved in the soil directly affects the mobility of
the herbicide.

Silva et al. (2018) verified the effect of the amount of sugar cane
straw on the soil surface on the leaching of aminocyclopyrachlor. The
results showed that the sugarcane straw did not prevent the aminocyclopyrachlor from reaching the soil and being detected in the entire profile of the soil column (0–30 cm); however, in the leached solution only
traces of the product were found (<0.21%). The authors concluded that in
greater amounts of sugarcane straw (20 t ha^{-1}), 40% more of the product
was retained in the straw; however, this effect can increase losses of the
herbicide by photodegradation and volatilization, consequently decreasing control of weeds.

3.7.3   Leaching of herbicides in lysimeters under field conditions

Studies of leaching of herbicides in a lysimeter show results based
on the percentage of the leached substance on the total applied at the
beginning of the study (Dörfler et al. 2006). Weber et al. (2006) studied the leaching potential of metolachlor, atrazine, and primisulfuron-
methyl in three fallow fields with soils of different physical-chemical
characteristics for 128 days after application. The results indicated
that the largest amounts of leachate herbicides occurred in increasing
order: metolachlor, primisulfuron-methyl, and atrazine. The correlation
analyses indicated that the leaching of the three herbicides was directly
related, with the values of Rf, DT$_{50}$, leached volume, S$_w$, and the average
percentage of OM.

Information on the leaching of herbicides in different cultivation systems can also be provided. Mikata et al. (2003) evaluated the
leaching of imazosulfuron in three lysimeters over the cultivation of
wheat, barley, and rapeseed over a 3-year period. A second dose of
imazosulfuron, 50 g a.i. ha^{-1}, was applied to one of the lysimeters, as
recommended for the culture. In addition, fertilizers and other pesticides were applied to assess the leaching of imazosulfuron on all agricultural practices adopted in the region, so that the herbicide and its
metabolites had little leaching potential when applied to these cropping systems.

Another way of using the method was described by Schroll and Kühn (2004). In addition to assessing the pesticide leaching, it allows detecting and quantifying ^{14}C losses through product volatilization and mineralization in the system. The authors obtained values of 0.28% to 0.68% of leaching of isoproturon in different soils in relation to the total applied. This system consists of two chambers, which are located on the surface of the soil and also on plants. Using this system, Grundmann et al. (2008) verified the leaching of isoproturon up to 48 days after application and glyphosate for 15 months and three sequential applications, in a dose of 1 kg a.i. ha^{-1}. The results indicated that only 1.3% of the isoproturon was leached and that 90% of glyphosate was found in the 5 cm topsoil; however, it was emphasized that despite the low leaching potential of both products, leaching can occur by preferential flow of water in the soil. Another process that influenced the leaching of herbicides was the high rate of mineralization.

All the methodologies presented previously for analyzing radiolabeled substances must present safe results in accordance with the guidelines (EPA 1998, 2008; OECD 2000, 2004), which determine the percentage of recovery of the studies. Table 3.3 presents additional results about mobility studies with ^{14}C-herbicides, indicating more possibilities and applications of the methodologies mentioned in this chapter.

3.8 Concluding remarks

In this chapter, detailed methodologies for mobility studies are presented through mobility in soil TLC, in soil columns, as well as leaching in the field via lysimeters, indicating that the use of radiolabeled ^{14}C molecules usually promotes advantages over other conventional studies. The main advantage is related to the possibility of accurately determining the minimum amounts of the herbicide in relatively short periods.

In the descriptions of the studies additional information is addressed, such as dimension and purpose, in order to support the choice of the best method to be used with ^{14}C-herbicide by the researchers. Furthermore, there are indications of the necessary information to be made available when reporting. Such information is important for all researchers who wish to use radiolabeled herbicides to characterize their transport, whether for the legal registration of products or for investigating the behavior of these products in the soil profile. Understanding this behavior allows the analysis of the risk of herbicides to the environment, and makes it possible to determine their proper use in chemical control of weeds for better efficacy.

Table 3.3 Mobility studies of herbicides in soil TLC, columns, and lysimeters

Herbicide transport methods	Applied amount of ^{14}C	Herbicide	Evaluation time	Results	Reference
Soil TLC	0.30 kBq	^{14}C-Hexazinone	30 min	Rf = 0.9155 (high mobility)	Mendes et al. (2019b)
	0.53 kBq	^{14}C-Metribuzin		Rf = 0.8805 (mobile)	
	0.36 kBq	^{14}C-Quinclorac		Rf = 0.9490 (high mobility)	
	185 MBq mg^{-1}	^{14}C-Diazinon	–	Rf = 0.05–0.23 (low mobility)	Arienzo et al. (1994)
	≅0.16 kBq	^{14}C-Ethofumesate	–	Rf = 0.38 (moderate mobility)	Sánchez-Martín et al. (1994)
	≅0.16 kBq	^{14}C-Metalochlor	–	Rf = 0.35 (moderate mobility)	
	≅0.83 kBq	^{14}C-Fluometuron	–	Rf = 0.48 (moderate mobility)	
	≅1.66 kBq	^{14}C-Atrazine	–	Rf = 0.47 (moderate mobility)	
		^{14}C-Glyphosate		Rf = 0.05 (low mobility)	
	4.29 MBq mg^{-1}	^{14}C-Atrazine	–	Soil surface, Rf = 0.10 (not mobile) Soil subsurface, Rf = 0.81 (high mobility)	Jacobsen et al. (2001)
Soil column	795.5 MBq mmol^{-1}	^{14}C-Prosulfocarb	–	Rf = 0.51 to 0.63 (moderate mobility)	Braun et al. (2017)
	369.26 kBq	^{14}C-Glyphosate	11 months	27.6% to 98.8% glyphosate e AMPA	Al-Rajab et al. (2008)
	16,666.67 MBq	^{14}C-Hexazinone ^{14}C-Metribuzin ^{14}C-Quinclorac	48 hours	0% to 41%	Mendes et al. (2019b)
	3,700 MBq	^{14}C-Paraquat	10 days	0% to 4% (present in leachate)	Ismail et al. (2013)
	3.16 MBq mg^{-1}	^{14}C-Prosulfocarb	1 and 28 days	4.21% to 32.5% (leachable) 43.3% to 85.2% (soil)	Barba et al. (2020)
	3.45 MBq	^{14}C-Mesotrione	48 hours	<1.45% to 87.41% (present in leachate)	Mendes et al. (2017)
	2.85 MBq mg^{-1}	^{14}C-Ethofumesate	24 hours	96.0 % (present in leachate) 3.35%–84.6% (present in leachate) *soil + OM	Marín-Benito et al. (2018)
	0.13–0.27 MBq l^{-1}	^{14}C-Glyphosate	96 hours	0.3% to 10.5% (present in leachate)	Jonge et al. (2000)

(Continued)

Table 3.3 Mobility studies of herbicides in soil TLC, columns, and lysimeters *(Continued)*

Herbicide transport methods	Applied amount of ^{14}C	Herbicide	Evaluation time	Results	Reference
Lysimeter	0.69 MBq mg⁻¹	¹⁴C-Isoproturon	4 years	<0.4% (present in leachate)	Dörfler et al. (2006)
	5.32 MBq	¹⁴C-Glyphosate	2 years	27% of glyphosate and AMPA (sand soil) 59% of glyphosate and AMPA (clay soil)	Bergström et al. (2011)
	0.56 MBq 0.57 MBq 0.57 MBq	¹⁴C-metolachlor ¹⁴C-atrazine ¹⁴Cprimisulfuron-methyl	128 days	42,3–64,8% (present in soil) 38.6–46.9% (present in soil) 30.7–58.8% (present in soil)	Weber et al. (2006)
	0.55 MBq 0.81 MBq 0.54 MBq	¹⁴C-metolachlor ¹⁴C-atrazine ¹⁴C-primisulfuron-methyl	128 days	20.86–48.90% (present in soil) 20.69–53.60% (present in soil) 15.78–41.55% (present in soil)	Weber et al. (2007)
	5.156 MBq	¹⁴C-Glyphosate	2 years	n.d.[a] (present in leachate) 97% of glyphosate and AMPA (present in soil)	Fomsgaard et al. (2003)
	3.4 MBq mg⁻¹	¹⁴C-Mefenacet	6 years	0.778% (present in leachate) 44.58% (present in soil)	Kyung et al. (2015)
	1.66 MBq mg⁻¹	¹⁴C-Cinosulfuron	4 years	2.43–2.99% (present in leachate) 56.71–57.52% (present in soil)	Lee et al. (2002)
	421.587 KBq mg⁻¹	¹⁴C-Atrazine	22 years	93% (present in soil)	Jablonowski et al. (2000)

[a] n.d., not detected

References

Al-Rajab, A. J., S. Amellal and M. Schiavon. 2008. Sorption and leaching of ^{14}C-glyphosate in agricultural soils. *Agronomy for Sustainable Development* 28:419–428.

Arienzo, M., T. Crisanto, M. J. Sanchez-Martin and M. Sanchez-Camazano. 1994. Effect of soil characteristics on adsorption and mobility of (^{14}C) diazinon. *Journal of Agricultural and Food Chemistry* 42:1803–1808.

Barba, V., J. M. Marín-Benito, M. J. Sánchez-Martín and M. S. Rodríguez-Cruz. 2020. Transport of ^{14}C-prosulfocarb through soil columns under different amendment, herbicide incubation and irrigation regimes. *Science of the Total Environment* 701:1–8.

Barriuso, E., C. Feller, R. Calvet and C. Cerri. 1992. Sorption of atrazine, terbutryn and 2,4-D herbicides in two Brazilian Oxisols. *Geoderma* 53:155–167.

Bergström, L., E. Börjesson and J. Stenström. 2011. Laboratory and lysimeter studies of glyphosate and aminomethylphosphonic acid in a sand and a clay soil. *Journal of Environmental Quality* 40:98–108.

Braun, K. E., A. K. Luks and B. Schmidt. 2017. Fate of the ^{14}C-labeled herbicide prosulfocarb in a soil and in a sediment-water system. *Journal of Environmental Science and Health Part B* 52:122–130.

Bukun, B., D. L. Shaner, S. J. Nissen, P. Westra and G. Brunk. 2010. Comparison of the interactions of aminopyralid vs. clopyralid with soil. *Weed Science* 58: 473–477.

Carter, A. D. 2000. Herbicide movement in soils: principles, pathways and processes. *Weed Research* 40:113–122.

Carvalho, D. D., L. D. Silva, J. G. Guerra, F. A. Cruz and A. P. Souza. 2007. Instalação, calibração e funcionamento de um lisímetro de pesagem. *Engenharia Agrícola* 27:363–372.

Christoffoleti, P. J., R. F. L. Ovejero, V. Damin, S. J. P. Carvalho and M. Nicolai. 2008. *Comportamento dos Herbicidas Aplicados ao Solo na Cultura da Cana-de-Açúcar.* Basf: Piracicaba.

Cornejo, J. and P. Jamet. 2000. *Pesticide/Soil Interactions: Some Current Research Methods.* Paris: Institut National De La Recherche Agronomique.

Deng, H., D. Feng, J. X. He, F. Z. Li, H. M. Yu and C. J. Ge. 2017. Influence of biochar amendments to soil on the mobility of atrazine using sorption–desorption and soil thin-layer chromatography. *Ecological Engineering* 99:381–390.

Diene, A. A. 2004. Desenvolvimento de tensiômetros para sucção elevada, ensaiados em lisímetros de laboratório. Dissertação de Mestrado, Programa de Pós-Graduação em Engenharia Civil, COPPE, Universidade Federal do Rio de Janeiro. http://www.dominiopublico.gov.br/pesquisa/DetalheObraForm .do?select_action=&co_obra=85820 (accessed January 02, 2020).

Dörfler, U., G. Cao, S. Grundmann and R. Schroll. 2006. Influence of a heavy rainfall event on the leaching of [^{14}C] isoproturon and its degradation products in outdoor lysimeters. *Environmental Pollution* 144:695–702.

Environmental Protection Agency (EPA). 1998. *Soil Thin Layer Chromatography.* EPA, 6 p. Washington: EPA – Fate transport and transformation test guidelines – OPPTS 835.1210.

Environmental Protection Agency (EPA). 2008. *Leaching Studies.* EPA, 16 p. Washington: EPA – Fate transport and transformation test guidelines – OPPTS 835.1240.

Food and Agriculture Organization of the United Nations (FAO). 1989. *Revised Guidelines on Environmental Criteria for the Registration of Pesticides.* 51 Rome: FAO.

Flessner, M. L., G. R. Wehtje, J. S., McElroy and J. A. Howe. 2015. Methiozolin sorption and mobility in sand-based root zones. *Pest Management Science* 71:1133–1140.

Fomsgaard, I. S., N. H. H. Spliid and G. Felding. 2003. Leaching of pesticides through normal-tillage and low-tillage soil—a lysimeter study. I. Isoproturon. *Journal of Environmental Science and Health Part B* 38:1–18.

Flury, M. 1996. Experimental evidence of transport of pesticides through field soils--a review. *Journal of Environmental Quality* 25:25–45.

Führ, F., W. Steffens, W. Mittelstaedt and B. Brumhard. 1991. Lysimeter experiments with ^{14}C-labelled pesticides–an agroecosystem approach. *Pesticide Chemistry* 14:37–47.

Grundmann, S., U. Dörfler, B. Ruth, C. Loos, T. Wagner, H. Karl, J. C. Munch and R. Schroll. 2008. Mineralization and transfer processes of ^{14}C-labeled pesticides in outdoor lysimeters. *Water, Air, & Soil Pollution* 8:177–185.

Guimarães, A. C. D., K. F. Mendes, T. F. Campion, P. J. Christoffoleti and V. L. Tornisielo. 2019. Leaching of herbicides commonly applied to sugarcane in five agricultural soils. *Planta Daninha* 37:1–9.

Guo, J., G. Zhu, J. Shi and J. Sun. 2003. Adsorption, desorption and mobility of fomesafen in Chinese soils. *Water, Air, & Soil Pollution* 148:77–85.

Haberhauer, G., B. Temmel and M. H. Gerzabek. 2002. Influence of dissolved humic substances on the leaching of MCPA in a soil column experiment. *Chemosphere* 46:495–499.

Helling, C. S. 1971. Pesticide mobility in soils I. Parameters of thin-layer chromatography. *Soil Science Society of America Journal* 35:732–737.

Helling, C. S. and B. C. Turner. 1968. Pesticide mobility: determination by soil thin-layer chromatography. *Science* 162:562–563.

Helling, C. S. and Dragun, J. Soil leaching tests for toxic organic chemicals. 1981. In: *Test Protocols for Environmental Fate and Movement of Toxicants: Proceedings of a symposium, Association of Official Analytical Chemists, 94th annual meeting, October 21, 22, 1980, Washington, DC.* p. 43–48. Washington: Association of Official Analytical Chemists.

Hunter, J. H. and E. H. Stobbe. 1972. Movement and persistence of picloram in soil. *Weed Science* 20:486–489.

Inoue, M. H., D. C. Santana, R. S. Oliveira Jr., R. A. Clemente, R. Dallacort, A. C. S. Possamai, C. T. C. Santana and K. M. Pereira. 2010. Potencial de lixiviação de herbicidas utilizados na cultura do algodão em colunas de solo. *Planta Daninha* 28:825–833.

Isensee, A. R. and A. M. Sadeghi. 1992. Laboratory apparatus for studying pesticide leaching in intact soil cores. *Chemosphere* 25:581–590.

Ismail, B. S., M. Sameni and M. Halimah. 2013. Comparison of the mobility of the herbicides 2, 4-D and ^{14}C-paraquat in selected Malaysian agricultural soils. *International Journal of Plant, Animal and Environmental Sciences* 3:1–9.

Jablonowski, N. D., S. Köppchen, D. Hofmann, A. Schäffer and P. Burauel. 2000. Persistence of ^{14}C-labeled atrazine and its residues in a field lysimeter soil after 22 years. *Environmental Pollution* 157:2126–2131.

Jacobsen, C. S., N. Shapir, L. O. Jensen, E. H. Jensen, R. K. Juhler, J. C. Streibig, R. T. Mandelbaum and A. Helweg. 2001. Bioavailability of triazine herbicides in a sandy soil profile. *Biology and Fertility of Soils* 33:501–506.

Jones, D. L., G. Edwards-Jones and D. V. Murphy. 2011. Biochar mediated alterations in herbicide breakdown and leaching in soil. *Soil Biology and Biochemistry* 43:804–813.

Jonge, H., L. S. W. Jonge and O. H. Jacobsen. 2000. ^{14}C-Glyphosate transport in undisturbed topsoil columns. *Pest Management Science* 56:909–915.

Katagi, T. 2013. Soil column leaching of pesticides. In: *Reviews of Environmental Contamination and Toxicology* ed. D. Whiteacre, Volume 221. New York: Springer.

Khan, S. U. 1980. Occurrence and persistence of pesticide residues in soil. In: *Pesticides in the Soil Environment*, ed. S. U. Khan, 163-198 p. Amsterdam: Elsevier.

Kördel, W., M. Herrchen and W. Klein. 1991. Experimental assessment of pesticide leaching using undisturbed lysimeters. *Pesticide Science* 31:337–348.

Koskinen, W. C., A. M. Cecchi, R. H. Dowdy and K. A. Norberg. 1999. Adsorption of selected pesticides on a rigid PVC lysimeter. *Journal of Environmental Quality* 28:732–734.

Kyung, K. S., K. C. Ahn, J. W. Kwon, Y. P. Lee, E. Y. Lee, Y. J. Kim and J. K. Lee. 2015. Long-term fate of the herbicide mefenacet in a rice-grown lysimeter over a period of 6 consecutive years. *Journal of the Korean Society for Applied Biological Chemistry* 58:35–43.

Laabs, V., W. Amelung, A. Pinto and W. Zech. 2002. Fate of pesticides in tropical soils of Brazil under field conditions. *Journal of Environmental Quality* 31:256–268.

Lee, J. K., F. Führ, J. W. Kwon and K. C. Ahn. 2002. Long-term fate of the herbicide cinosulfuron in lysimeters planted with rice over four consecutive years. *Chemosphere* 49:173–181.

Liu, X., H. Wu, T. Hu, X. Chen and X. Ding. 2018. Adsorption and leaching of novel fungicide pyraoxystrobin on soils by ^{14}C tracing method. *Environmental Monitoring and Assessment* 190:86.

Luchini, L. and M. M. Andréa. 2018. The use of nuclear techniques for environmental studies. In: *Integrated Analytical Approaches for Pesticide Management*, ed. B. Maestroni and A. Cannavan, 165–182. Cambridge: Academic Press.

Marín-Benito, J. M., M. J. Sánchez-Martín, J. M. Ordax, K. Draoui, H. Azejjel and M. S. Rodríguez-Cruz. 2018. Organic sorbents as barriers to decrease the mobility of herbicides in soils. Modelling of the leaching process. *Geoderma* 313:205–216.

Mendes, K. F., B. A. B. Martins, F. C. Reis, A. C. R. Dias and V. L. Tornisielo. 2017. Methodologies to study the behavior of herbicides on plants and the soil using radioisotopes. *Planta Daninha* 35:1–21.

Mendes, K. F., G. P. Olivatto, R. N. Sousa, L. V. Junqueira and V. L. Tornisielo. 2019a. Natural biochar effect on sorption–desorption and mobility of diclosulam and pendimethalin in soil. *Geoderma* 347:118–125.

Mendes, K. F., K. E. Hall, V. Takeshita, M. L. Rossi and V. L. Tornisielo. 2018. Animal bonechar increases sorption and decreases leaching potential of aminocyclopyrachlor and mesotrione in a tropical soil. *Geoderma* 316:11–18.

Mendes, K. F., R. N. Sousa, V. Takeshita, F. G. Alonso, A. P. J. Régo and V. L. Tornisielo. 2019b. Cow bone char as a sorbent to increase sorption and decrease mobility of hexazinone, metribuzin, and quinclorac in soil. *Geoderma* 343:40–49.

Mikata, K., F. Schnöder, C. Braunwarth, K. Ohta and S. Tashiro. 2003. Mobility and degradation of the herbicide imazosulfuron in lysimeters under field conditions. *Journal of Agricultural and Food Chemistry* 51:177–182.

Monquero, P. A., P. V. Silva, A. C. Silva-Hirata, D. C. Tablas and I. Orzari. 2010. Lixiviação e persistência dos herbicidas sulfentrazone e imazapic. *Planta Daninha* 28:185–195.

Organisation for Economic Co-operation and Development (OECD). 2000. *Guidance Document for the Performance of Out-Door Monolith Lysimeter Studies.* OECD Environmental Health and Safety Publications, 25 p. Paris: OECD (Test 106).

Organisation for Economic Co-operation and Development (OECD). 2004. *OECD Guidelines for the Testing of Chemicals – Leaching in Soil Columns.* OECD Environmental Health and Safety Publications, 15 p. Paris: OECD (Test 312).

Oliveira, M. F. 2011. Comportamento de Herbicidas no Ambiente. In: *Plantas Daninhas e seu Manejo*, ed. R. S. Oliveira Jr. and J. Constantin, 315–362. Guaíba: Agropecuária.

Oliveira, M. F. and A. M. Brighenti. 2011. Comportamento de herbicidas no ambiente. In: *Biologia e Manejo de Plantas Daninhas*, ed. R. S. Oliveira Jr., M. H. Inoue and J. Constantin, 263–304. Curitiba: Omnipax Editora.

Pimentel, D. 1995. Amounts of pesticides reaching target pests: environmental impacts and ethics. *Journal of Agricultural and Environmental Ethics* 8:17–29.

Pesticide Properties Data Base (PPDB). 2020. Glyphosate (Ref: MON 0573). https://sitem.herts.ac.uk/aeru/footprint/es/Reports/373.htm (accessed January 02, 2020).

Ravanel, P., M. Liégeois, D. Chevallier and M. Tissut. 1999. Soil thin-layer chromatography and pesticide mobility through soil microstructures. *Journal of Chromatography A* 864:145–154.

Sánchez-Martín, M. J., T. Crisanto, M. Arienzo and M. Sánchez-Camazano. 1994. Evaluation of the mobility of ^{14}C-labelled pesticides in soils by thin layer chromatography using a linear analyser. *Journal of Environmental Science & Health Part B* 29:473–484.

Schroll, R, H. H. Becher, U. Dörfler, S. Gayler, S. Grundmann, H. P. Hartmann and J. Ruoss. 2006. Quantifying the effect of soil moisture on the aerobic microbial mineralization of selected pesticides in different soils. *Environmental Science & Technology* 40:3305–3312.

Schroll, R. and S. Kühn. 2004. Test system to establish mass balances for ^{14}C-labeled substances in soil–plant–atmosphere systems under field conditions. *Environmental Science & Technology* 38:1537–1544.

Sherma, J. and B. Fried. 2003. *Handbook of Thin-Layer Chromatography*. Boca Raton: CRC Press/Taylor & Francis Group.

Sherma, J. and F. Rabel. 2018. A review of thin layer chromatography methods for determination of authenticity of foods and dietary supplements. *Journal of Liquid Chromatography & Related Technologies* 41:645–657.

Si, Y., S. Wang, J. Zhou, R. Hua and D. Zhou. 2005. Leaching and degradation of ethametsulfuron-methyl in soil. *Chemosphere* 60:601–609.

Silva, G. S., A. F. M. Silva, K. F. Mendes, R. F. Pimpinato and V. L. Tornisielo. 2018. Influence of sugarcane straw on aminocyclopyrachlor leaching in a greencane harvesting system. *Water, Air & Soil Pollution* 229:156–163.

Silva, L. L., C. L. Donnici and J. D. Ayala. 2009. Traçadores: o uso de agentes
    químicos para estudos hidrológicos, ambientais, petroquímicos e biológicos.
    Química Nova 32:1576–1585.
Strebe, T. A. and R. E. Talbert. 2001. Sorption and mobility of flumetsulam in
    several soils. *Weed Science* 49:806–813.
Weber, J. B., K. A. Taylor and G. G. Wilkerson. 2006. Soil and herbicide properties
    influenced mobility of atrazine, metolachlor, and primisulfuron-methyl in
    field lysimeters. *Agronomy Journal* 98:8–18.
Weber, J. B., R. L. Warren, L. R. Swain and F. H. Yelverton. 2007. Physicochemical
    property effects of three herbicides and three soils on herbicide mobility in
    field lysimeters. *Crop Protection* (3)26:299–311.

chapter four

Anaerobic and aerobic degradation studies of herbicides and radiorespirometry of microbial activity in soil

Kassio Ferreira Mendes[1], *Kamila Cabral Mielke*[1],
Lucas Heringer Barcellos Júnior[1], *Ricardo Alcántara de la Cruz*[2], *Rodrigo Nogueira de Sousa*[3]
[1]*Department of Agronomy, Federal University of Viçosa*
[2]*Department of Chemistry, Federal University of São Carlos*
[3]*Department of Soil Science, College of Agriculture University of São Paulo*

Contents

4.0   Introduction

Weeds are great competitors with cultures for resources of the environment fundamental to their growth and physiological development, such as water, nutrients, light, and space. These plants, when present in agricultural areas, may cause great damage to farmers, and for this reason, herbicides are the most consumed pesticides around the world (FAO 2019).

Herbicides can be applied directly to the soil (in pre-emergence) or on the target weed (in post-emergence). In pre-emergence applications, the herbicide is in direct contact with the soil, and in post-emergence only 30% to 40% of the product reach the target, and a large part is directed to the soil (Law 2001). In both situations there is a high chance of soil contamination, which may have a negative impact on groundwater.

The behavior of the herbicide in the soil is governed by the physico-chemical properties of the molecule and the soil, and can have retention, transport, and transformation processes (Arias-Estévez et al. 2008). The process of transformation the molecule is a function of its degradation into secondary compounds (metabolites), which can be through physical, chemical, and biological processes.

Biological degradation is the most common way to dissipate the herbicide in the environment, and occurs through the soil microbiota (bacteria, fungi, protozoa, and actinomycetes), which uses the herbicidal molecule as an energy source and transforms it into compounds without herbicidal action, a process also known as detoxification (Maier 2000). In addition, the degradation of herbicides may be due to co-metabolism, in which the microorganism does not obtain energy or benefit from the degradation. The transformation process is usually mediated by non-specific enzymes that are capable to transform various organic compounds (Reis et al. 2019). The complexity of the molecule influences the higher or lower facility of microorganisms to degrade it, characterizing it in low or high persistence in the soil (Bending et al. 2006), being measured by the molecule's half-life (degradation time $[DT]_{50}$).

The degradation of the herbicide in the soil by microorganisms can occur under aerobic or anaerobic conditions. In the presence of oxygen, the herbicide is mineralized in CO_2 and water, and in CH_4, CO_2, and water in conditions without oxygen (OECD 2002a). The efficiency of the degradation of organic compounds, such as herbicides, by aerobic bacteria is higher than that of anaerobic bacteria, as they use oxygen as an oxidizing agent, and they are present in the region of the soil where there is a higher content of organic matter and an excellent soil-water-air ratio for the microbiota (Gebler and Spadoto 2008). In conditions of absence of oxygen, the herbicide can become more persistent in the soil, and its degradation pathways are different from microorganisms with aerobic metabolism (Wang et al. 2014).

The respiration of microorganisms in the soil can be affected in the presence of herbicides, depending on the dose and persistence of the herbicide and soil attributes, especially of organic matter and clay content. Reis et al. (2019) observed greater microbial respiration rate in sandy soil than in clayey soil, submitted to the application of diuron, at 42 days after treatment. The sandy soil had a lower organic matter and clay content, favoring the bioavailability of this herbicide in the soil solution, and consequently being more exposed to microorganisms (Ahtiainen et al. 2003). The higher respiration of the microbial population was probably due to the herbicide being a source of energy and carbon for the microorganisms to multiply, and consequently to increase the microbial activity in the soil (Imfeld and Vuilleumier 2012).

There are some techniques for quantifying the degradation of herbicides in soils. Among these, the use of radiolabeled carbon molecules (^{14}C) stands out, as it is an analytical tool of high sensitivity, precision, specificity, and rapid analysis (Luchini and Andréa 2018). This separation method, when combined with others, such as chromatography, allows for high selectivity, sensitivity, and quantitative detection of herbicides and their metabolites (Wang et al. 2014), making it possible to efficiently measure molecule residues in the environment.

This technique makes it possible to determine the relative ability of microorganisms to completely mineralize ^{14}C-herbicides into ^{14}C-CO_2 by studying radiorespirometry with ^{14}C-glucose (Posen et al. 2006), in addition to studies of biological degradation and mineralization of ^{14}C-herbicides in soil (Mueller and Senseman 2015). These studies are performed in the laboratory, and the biometric flask, popularly known as Bartha and Pramer flasks (Bartha and Pramer 1965), is commonly used. This flask is a 250-mL Erlenmeyer fused to a 50-mL test tube. Usually, 50 g of soil is added to the flask and maintained at 70% of the field capacity. As the microbial population breathes, the release of CO_2 that is captured by a strong base, for example, KOH or NaOH, increases, being quantified later by liquid scintillation spectrometer (LSS) (Bartha and Pramer 1965).

The degradation and respiration microbial biomass studies are important in the knowledge of the management that will be adopted in the agricultural system. It is necessary for farmers and technicians to plan the crops that will be rotated in the next harvest seasons in order to establish strategies for the control of weeds. These studies indicate the persistence of the herbicide in the soil, which implies the residual period of the product for the control of weeds, injuries of the crop in succession (carryover), and the soil and water contamination risk. Furthermore, these studies are mandatory for herbicide registration. In Brazil, studies in three soil classes are required (Latossolo Vermelho Escuro, clay loam texture; Latossolo Roxo, clay texture; and Gleissolo Húmico, sandy loam) (IBAMA 2012).

This chapter will approach the most relevant issues in studies of herbicide degradation by microorganisms under aerobic and anaerobic conditions, and microbial respiration in the soil, using the resources of ^{14}C radiometric techniques. It is hoped that this information will make it possible to conduct research involving quantifying the fate of

¹⁴C-herbicides in the soil, and to guide field technicians to make safer recommendations from an agronomic and environmental point of view.

4.1   Concepts

For a better understanding of the studies of anaerobic and aerobic degradation of herbicides and radiorespirometry of activity in soil are illustrated at Figure 4.1 and described below the main concepts based on the international guidelines of OECD (2000 and 2002a):

Persistence: Time of permanence of herbicide in the soil expressed in unit of time by the quantified concentration of the product, and the herbicide may be active or inactive in the soil. Herbicide residues can be quantified in the order of μg and ng kg⁻¹ of soil. The persistence of herbicide is always greater than its residual effect.

Residual effect: Period in which the herbicide has pre-emergence weed control, which is biologically active in the soil solution.

Carryover: Phytotoxic effect of the herbicide on the rotation or succession of crops.

Dissipation: Disappearance of the herbicide from the site of application in the soil through the processes of retention, transformation, and transport.

Aerobic degradation: Herbicide transformation reactions into metabolites and/or minerals in the presence of oxygen as a result of

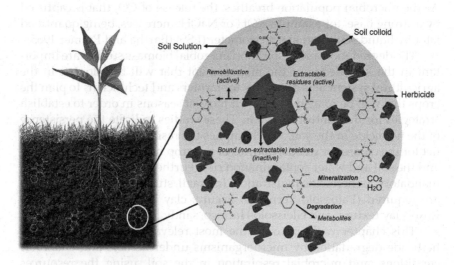

Figure 4.1 Schematization of the different herbicide degradation processes in the soil.

chemical and biological processes (e.g., hydrolysis, photolysis, microbial metabolism).

Anaerobic degradation: Herbicide transformation reactions into metabolites and/or minerals in the absence of oxygen.

Metabolites: All subproducts resulting from incomplete degradation of the herbicide that are available in the soil solution or as bound residues in soil colloid.

Mineralization: Products of the complete degradation of herbicide or metabolites in the form of minerals, mainly evolved CO_2 and H_2O under aerobic conditions in the soil. A complete degradation of herbicides under anaerobic conditions implies the formation of CO_2, H_2O, and CH_4. In this study, when ^{14}C-herbicides are used, mineralization means extensive degradation during which a labeled carbon atom is oxidized with release of the appropriate amount of ^{14}C-CO_2.

Extractable residues: Residues of herbicides and/or metabolites extracted from soil with solvents, which are bioavailable (active) in the soil.

Bound (non-extractable) residues: Residues of herbicides and/or metabolites not extracted from soil with solvents, which are unavailable (inactive) in soil solution, because they are bound to soil colloids and remain persisting for time.

Remobilization: Return of the herbicide bonded residue and/or metabolite from the solid phase of the soil to the soil solution (liquid phase) making it active.

Solid phase: Mineral and organic matter (plant and animal remain). The organic fraction is undergoing a continuous renovation process but is reduced after the introduction of agricultural systems.

Liquid phase. Mixture of water, mineral salts and low molecular weight molecules, such as amino acids, peptides, sugars, and humic substances. Most of the herbicide molecules are deposited in this fraction of the soil and are subject to numerous reactions, depending on the properties of the solution (pH, ionic strength, and redox potential [Eh]), which determines their fate in the environment.

Mass balance: Sum of all the results of the amounts of the herbicide found in the soil as bound residue, extractable residue, and mineralized (percentage of radioactivity in relation to the total applied) obtained, in order to verify the consistency of the results and the conservation of mass in relation to that applied initially in the studied system.

Radiorespirometry: Evaluation of respiratory activity of microorganisms with the presence of ^{14}C-glucose as an energy source.

Half-life time: Time required to degrade or mineralize 50% of the herbicide initially applied can be described by first-order kinetics. known as DT_{50} (degradation time) or MT_{50} (mineralization time), respectively, in a study conducted in the laboratory, it is independent of the concentration. In field conditions, DT_{50} (dissipation time)

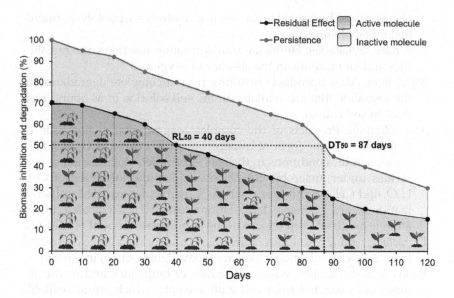

Figure 4.2 Schematic representation of residual effect (RL_{50}) and degradation time (DT_{50}) of diclosulam herbicide. (Adapted from Lavorenti et al. 2003, Dan et al. 2011.)

is determined, and it is the time required to dissipate 50% of the herbicide. It is also common to find DT_{80} and DT_{90}, corresponding to 80% and 90%, respectively, of herbicide dissipation or degradation.

Biometric flask: Hermetically sealed flask commonly used in biodegradation and respirometry studies of microorganisms using radioisotopes (^{14}C-herbicides).

RL_{50}: Herbicide residue level that inhibits 50% of the biomass of product-sensitive plants, known as bioindicator species. The difference between the values of RL_{50} and DT_{50} is shown in Figure 4.2.

Thin layer chromatography (TLC): These are TLC plates used for degradation studies with eluents, separating the herbicide from its metabolites.

Rf: Retention factor is the relationship between the distance traveled from the herbicide and the eluent using TLC.

4.2   Terrestrial field dissipation

The study of herbicide dissipation under field conditions is very important, as it demonstrates the actual condition of the study, evidencing the processes of transformation, retention, and transport of the product in the environment. However, when radiolabeled herbicides are taken out of the laboratory into the field, two restraints are immediately encountered – one regulatory, and one economic, according to Harvey Jr. (1983). First, state and federal nuclear regulatory agencies do not sanction the dispersion of radioactive materials (even ^{14}C) in uncontained or large-scale field experiments. Second, radiolabeled herbicides are expensive to purchase and synthesize, and their use in large-scale experiments is prohibitively expensive. Many herbicide field-dissipation studies, therefore, dispense with the radiolabel and utilize the normal formulated commercial herbicide instead (Harvey Jr. 1983). The conductance of these studies is directed and guided under very specific and rigorous conditions (U.S. EPA 2008a, Mueller and Senseman 2015).

Harvey Jr. (1983) reported that to conduct a field soil dissipation study which would (1) utilize radiotracer techniques; (2) employ simple, easily transported equipment; (3) could be installed in any cultivated, undisturbed, or even turf-covered soil; and (4) be capable of defining rate of breakdown, recovery of metabolites, and leaching process by lysimeter. A detailed description of the soil cylinder test placed in the field is described by Harvey Jr. (1983).

4.3 Biological degradation studies in biometric flasks

Evaluating the behavior of herbicides in the soil due to their biological degradation studies can also be performed through radiometric techniques in the laboratory. The method was established by the OECD 307 "Aerobic and Anaerobic Transformation in Soil" (OECD 2002a), which determines standards for the procedure to be performed. The method consists of measuring the mineralization rates of the ^{14}C-herbicide into ^{14}C-CO_2, in the extraction of the herbicide (extractable residue) and final soil burning (residue not extractable or bonded residue). The whole procedure should be performed under controlled conditions of temperature and humidity.

The study is applicable for nonvolatile, soluble, and water-insoluble compounds. For volatile products in the soil, the study does not apply, as they cannot be maintained under the conditions of this study.

4.4 Installation and conduction of the biological degradation study

Each experimental unit, composed of a biometric flask, popularly known as a flask of Bartha and Pramer (1965) (see Figure 4.3), containing 50 g of soil (dry base), must be incubated for a period between 2 and 28 days to $20 \pm 2°C$ (pre-incubation). The biometric flask of Anderson (1975) has also been widely used in herbicide degradation studies. The study should be

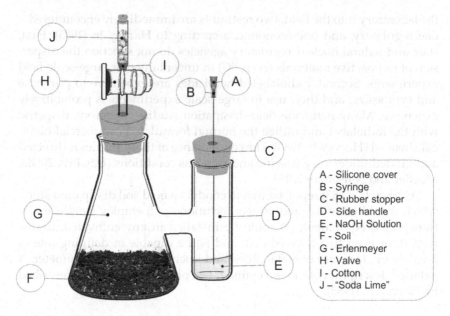

Figure 4.3 Schematic model of the biometric flask used in the aerobic degradation and mineralization studies of herbicides in soil (50 g dry base), modified from Bartha and Pramer (1965). (From Takeshita et al. 2019.)

carried out containing at least two replicates per sample only, in order to minimize the residues generated by the use of radiometric techniques. Before installation of the study, the soil moisture should be adjusted to 75% of the field capacity, and monitored during the incubation period of the study, where, if the difference between the initial water content and the given content is greater than 5%, the adjustment should be made adding distilled water.

The herbicide volume applied per vial is usually 200 µL of stock solution around 16.7 kBq (1,000,000 dpm), considering 50 g of soil per experimental unit. The solution should be applied with the aid of micropipette, and after the pre-incubation days must be homogenized in the soil with the aid of a glass rod, and the bottle should be closed. Subsequently, 10 mL of a sodium hydroxide solution (NaOH 0.1 mol L^{-1}) is added to the bottle handle and then the vials should be incubated at 20 ± 2°C or according to the focus temperature of the study.

4.5 Mineralization measurement for $^{14}C\text{-}CO_2$

Measurements of herbicide mineralization should be performed at intervals of days during the experimental period required for each herbicide (considering the mineralization time half-life – MT_{50}). In this evaluation, two aliquots are removed from the NaOH solution of each biometric vial, transferring them to scintillation vials, and the radioactivity of the evolved ^{14}C-CO_2 is determined in an LSS counter. The remaining solution should be replaced by a new one in each biometric vial. Final data are generated by ^{14}C-CO_2 contents accumulated during the course of the samplings.

4.6 ^{14}C-Extractable residue analysis (parental product and metabolites)

The extraction of herbicide residue from the soil is carried out with several solvents (e.g., methanol, water, acetone, toluene, among others), depending on the solubility of each herbicide studied. After the concentration of the extractable solution, it should be applied to TLC plates for the separation of parental product and metabolites formed in the biological degradation process. Concentrated extract aliquots and standard radiolabeled herbicide solution should be applied to silica plates, known as TLC plates. Elution should be conducted with solvents suitable for each herbicide in chromatographic tanks (vat). Products and possible metabolites can be observed with the help of ultraviolet light, x-ray film, or radio scanner, for the identification of peaks in chromatograms. The amount of each product (parental and metabolites) of concentrated extractable residues is evaluated using the retention factor (Rf) values, as described in TLC soil mobility studies. This technique does not allow identifying radiolabeled metabolites but allows identifying if there was the formation of radiolabeled metabolites. For specific identification of metabolites of samples, it is necessary to compare the chromatographic peaks with patterns of radiolabeled metabolites.

4.7 ^{14}C-Bound (non-extractable) residue analysis

The formation of bound residue should be verified by means of destructive samples, incubated in the same period and experimental conditions from which they are submitted to extraction with organic solvents, and the soil should finally be oxidized in a biological oxidizer, where the ^{14}C-CO_2 released during burning is trapped in a radioactive dioxide absorber solution, which is mixed with scintillation solution and quantified by LSS.

At the end of the study, the mass balance (recovery rate of the source product in %) should range from 90% to 110%, considering the total of ^{14}C-herbicide applied, mineralized in the form of ^{14}C-CO_2, extractable residue, and the bound residue herbicide in the soil (OECD 2002a).

4.8   Degradation and mineralization model

The result of the above studies is a series of mathematical expressions describing the relationship between rate of degradation and the variables of climate and initial concentration of herbicide (Laskowski et al. 1983). Mineralization data of ^{14}C-CO_2 accumulated and quantity of the degraded ^{14}C-herbicide should be adjusted for a first-order kinetics model (the most used model in this type of study), adapted to U.S. EPA (2020):

$$C_t = C_0 e^{-kt}$$

Where

- C_t = the concentration of the herbicide remaining in the soil according to time (%)
- C_0 = the concentration of the herbicide in zero time
- k = the speed of mineralization or constant degradation of decline 1/day
- t = the incubation time (days)

   From the k values, it is possible to calculate the half-life time of mineralization or degradation (MT_{50} or DT_{50}, respectively) of the herbicide. The MT_{50} or DT_{50} is defined as the time required for 50% of the herbicide applied to be mineralized or degraded, respectively, and should be calculated using the following equation:

$$MT_{50} \text{ or } DT_{50} = \frac{\ln 2}{k}$$

   For found DT_{90} value (90% of the herbicide applied to be degraded) use $\ln 10$. There are other models that can be adjusted, such as the second-order exponential degradation model:

$$C_t = C_1 e^{-k_1 t} + C_2 e^{-k_2 t}$$

In which

- C_t = the herbicide concentrated in the soil, in time
- C_1 and C_2 = the initial concentrations of the herbicide in the soil
- t = the incubation time (days)
- k_1 and k_2 = the dissipation speed constants of each phase ($k_1 > k_2$)

   In a biphasic model, degradation rates are usually much slower during the second stage; therefore, it is relevant to determine the time required for the degradation of 50% and 90% of the initial concentration, DT_{50} and DT_{90}, respectively (Oliveira Jr. et al. 2013).

U.S. EPA (2020) also described the *nth*-order rate model or indeterminate order rate equation model and double first-order in parallel for herbicide degradation kinetics equations.

In the end, the results will allow an estimate of the herbicide processing rate and also of the rate of formation and decline of metabolites under field conditions, and then this study provides information on the persistence of the herbicide applied to the soil, structurally altered by physical, chemical, and biological reactions (OECD 2002a).

4.9 Radiorespirometry of microbial activity in soil

Microbial activity influences many processes related to the transformation of inorganic molecules, decomposition of organic matter, suppression of pathogens, degradation of pesticides (including herbicides), phytohormone production, and bioremediation (Majewsky et al. 2010, Saleem and Moe 2014, Khatoon et al. 2017). Usually the pesticide degradation process is mediated by microorganisms capable of converting these substances into metabolites, CO_2, and H_2O. Pesticides once in the soil can be metabolized as a source of carbon and energy by heterotrophic microorganisms (OECD 2002b). However, herbicides applied in the soil can positively or negatively influence soil biological characteristics, e.g., reduction of microbial biomass and increase in respiratory rate, reduction or increase of both (Reis et al. 2008), poisoning of the microbiota, or even not causing significant effects (Mahía et al. 2008, Blume and Reichert 2015). The respiratory rate related to catabolism of herbicides in the soil can be monitored through the evolution of CO_2 produced from microbial respiration, being possible to evaluate the possible impacts on soil microbiota (Reid et al. 2001).

The application of radiorespirometry is a quick way to obtain information about the catabolic mechanisms operating in a biological system, being an effective research tool to study the effect of herbicides on soil microbial activity. Radiorespirometry describes the respiratory activity of a given biological system by analyzing kinetic data from the production of ^{14}C-CO_2 from carbon atoms of substrates individually labeled with ^{14}C (Wang 1963). The model substrate used to evaluate soil microbial activity is ^{14}C-glucose (Sheehan 1997), conferred by its characteristics of high solubility in water, easy degradation and bioavailability, and lower adsorption in the soil (Jones and Edwards 1998, Nguyen and Guckert 2001).

For the evaluation of radiorespirometry of a herbicide-treated soil, at least two test concentrations are recommended, which should be chosen in relation to the highest dose of herbicide recommended under field conditions, according to OECD 217 "Soil Microorganisms: Carbon

Transformation Test" (OECD 2000). Before the soil is used for the test, the moisture content should be adjusted to a value between 40% and 60% of the maximum water retention capacity and then pre-incubated for a period between 2 and 28 days at room temperature ($20 \pm 2°C$). This procedure allows the germination and removal of seeds, restoring the balance of microbial metabolism after changing sampling or storage conditions to incubation conditions (U.S. EPA 2008b).

The test is performed with about 50 to 200 g of soil (dry weight), commonly in biometric vials, under dark conditions and constant room temperature ($20 \pm 2°C$). An aliquot of standard ^{14}C-glucose analytical solution with determined specific activity is added to soil samples from each biometric vial at 0, 7, 14, and 28 days after herbicide application. If at the end of 28 days the differences between treated and untreated soils is equal to or greater than 25%, measurements will continue at 14-day intervals for a maximum period of 100 days (OECD 2000).

Bottles with treated soil are attached to the system, which can be continuous or closed airflow (Figures 4.4, 4.5, and 4.6). In a closed microbial respiration system (Figure 4.4), the soil sample (50 to 200 g) treated with ^{14}C-glucose and herbicide (non-radiolabeled) are transferred to the respirometric vials, for example biometric flask (Bartha and Pramer 1965), equipped with a side tube containing NaOH (0.2 mol L^{-1}). After the soil incubation period, aliquots of 1 mL are collected from the NaOH solution of each lateral loop of the biometric flask with the aid of a syringe, transferred to scintillation vials containing 10 mL of the scintillation

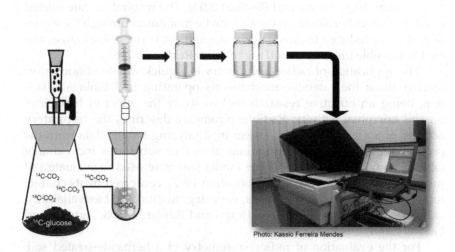

Photo: Kassio Ferreira Mendes

Figure 4.4 Schematic representation of the study of radiorespirometry in soil treated with herbicide in a biometric vial (Bartha and Pramer 1965) with closed airflow for the determination of ^{14}C-CO_2 evolved from ^{14}C-glucose added in soil.

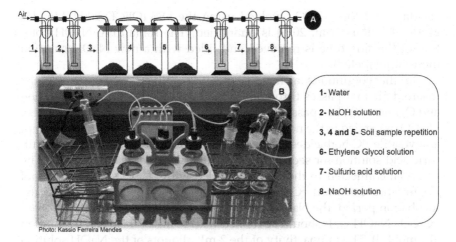

Figure 4.5 Schematic representation for each repetition (A) and equipment (B) of the study of radiorespirometry in soil treated with herbicide in a bottle with continuous airflow for the determination of ^{14}C-CO_2 evolved from ^{14}C-glucose added in soil.

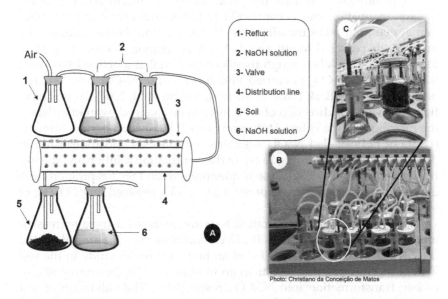

Figure 4.6 Schematic representation for treatment (A), equipment (B), experimental collection unit (C) of the study of radiorespirometry in soil treated with herbicide in a bottle with continuous airflow for the determination of ^{14}C-CO_2 evolved from ^{14}C-glucose added in soil.

cocktail, and the ^{14}C-CO$_2$ evolved using the LSS (OECD 2000, Mendes et al. 2018, Reis et al. 2019) is quantified. The remaining NaOH solution of the side tube is removed and filled with a new NaOH solution (non-radiolabeled).

In the continuous airflow system exemplified in Figure 4.5, the inserted air first passes through a NaOH solution for CO$_2$ sequestration and O$_2$ release. When passing through the vial with soil sample under test, ^{14}C-CO$_2$ is carried to the next vial with ethylene glycol solution to intercept any volatile ^{14}C-herbicides. The ethylene glycol vial is connected to a sulfuric acid solution for sequestration of alkaline volatile compounds. The next vial connected to the system is filled with NaOH solution (0.2 mol L^{-1}) to capture evolved ^{14}C-CO$_2$ resulting from microbial activity. After the incubation period, the NaOH solution is removed for analysis of evolved ^{14}C-CO$_2$. NaOH vials should be refilled with 10 mL of fresh NaOH solution (0.2 mol L^{-1}). The radioactivity of the 2-mL aliquots of the NaOH solution is measured by LSS after the addition of 10 mL of scintillation cocktail and a stabilization period of 24 h. After the final incubation period, the total amount of volatile ^{14}C-herbicides adsorbed by ethylene glycol is measured using 1 mL of the solutions and quantified by LSS by adding 10 mL of scintillation cocktail and maintained in a stabilization period of 1 h (U.S. EPA 2008b, Cycoń et al. 2011, Ismail et al. 2015).

In the other system with continuous airflow represented in Figure 4.6, the soil incubation vial is connected to the aeration tube of a distribution system that takes the air to the ^{14}C-CO$_2$ capture bottle. Initially, the inserted air passes through a vial of NaOH solution (0.4 mol L^{-1}). Then the air stream travels through the incubation vial at a defined flow rate. Subsequently, ^{14}C-CO$_2$ is carried and fixed to the NaOH solution connected to the soil flask (Wolf et al. 1994, Vivian et al. 2006). At desirable time intervals, the direction of air flows can be changed by manipulating the distribution line valve. At the end of each incubation period, aliquots (1 mL) of radiolabeled NaOH solutions are collected and transferred in duplicate to vials containing 10 mL of the scintillation solution and the initial concentration of ^{14}C-glucose is quantified in an LSS. Respiration rates induced by ^{14}C-glucose are expressed as ^{14}C-CO$_2$ released (mg CO$_2$/kg of soil/h).

The results of the concentration tests are analyzed using a regression model and the values of the effective concentration (EC) are calculated (EC$_{50}$, EC$_{25}$, and/or EC$_{10}$). The EC of an herbicide under study in the soil is the concentration that results in an inhibition of 50, 25, and/or 10% of carbon transformation into ^{14}C-CO$_2$, respectively. The validation of test results is based on mean differences of 25% between the ^{14}C-CO$_2$ released by the control treatment without herbicide application and the treated soil samples, so that large variations in controls can lead to false results (OECD 2000).

4.10   Examples of degradation studies of herbicides in soil

Several studies have been carried out in the world in order to evaluate the aerobic and anaerobic degradation of ^{14}C-herbicides, in order to enable the understanding in the transformation of these compounds in the soil and to mitigate the negative effects of environmental contamination.

Dictor et al. (2008) evaluated the degradation of ^{14}C-acetochlor in Neoluvisol and Calcosol. The rapid rate of degradation and the shorter degradation time of the herbicide were mainly attributed to the soil pH and organic carbon content. Calcosol has a more alkaline pH, and this variation in pH could have an effect on the availability of ^{14}C-acetochlor to microorganisms. According to these authors, the sorption of this herbicide is lower at more alkaline pH. Another parameter that may have affected the degradation of ^{14}C-acetochlor is the higher organic carbon content in Calcosol compared to Neoluvisol, which enabled a positive effect on the degradation activity of the soil microbiota.

The herbicide degradation is related to the dose applied to the soil. At the highest dose, ^{14}C-prosulfocarb showed a faster rate of degradation in the soil. However, the time required to degrade 50% of the molecule was longer compared to the lowest dose. This explanation is related to the lag phase involved in the degradation processes. The lag phase is the period of time that the microorganisms have to adapt to the new medium in the presence of the herbicide. During this period, cell multiplication may not occur, but there is a synthesis of new enzymes that may be necessary for the metabolization of the herbicide present in the soil. Soon after this period, microbial growth occurs quickly and exponentially. With the highest concentration of the herbicide in the soil, ^{14}C-prosulfocarb was more readily available to microorganisms, accelerating their degradation. On the other hand, the higher concentration of this herbicide required more time to be degraded (Barba et al. 2019).

The degradation of ^{14}C-quinclorac has been reported in five tropical soils with different physical and chemical attributes. The DT_{50} of this product varied between soils, but in all of them, the rate of degradation was very slow, and consequently, presented high DT_{50} values. Under the conditions of this study, pH was the main attribute that explained the results obtained. ^{14}C-quinclorac is a weak acid (pKa = 4.3), and at a pH higher than pKa, this herbicide is found in the soil solution in its dissociated form. In this situation, the degradation of ^{14}C-quinclorac is faster, as observed in this study (Alonso et al. 2019).

[14]C-metsulfuron-methyl has been studied in two soils in the United States. Interestingly, the highest rate of degradation and the lowest DT_{50} of this herbicide was found in Lithic Xeric Haplocambids, which had the lowest clay and organic carbon content (Trabue et al. 2006). The authors

reported that this occurs because ^{14}C-metsulfuron-methyl is largely dissipated via sorption kinetics, and soils with a high clay content have been shown to increase the sorption of this herbicide, making it less susceptible to soil microorganisms.

The fate of herbicides in the soil is greatly affected by the depth at which the compost is found in the soil. The ^{14}C-isoproturon had DT_{50} ranging from 10 to 990 days in the superficial and subsurface layer, respectively (Alletto et al. 2006). These authors observed that the main reason for this was the reduction of the organic carbon content in deeper layers in all evaluated soils. The low organic carbon content resulted in fewer microorganisms, and consequently increased the DT_{50} of ^{14}C-isoproturon.

Five herbicides had their degradation evaluated in a soil in Toulouse, France. ^{14}C-glyphosate, ^{14}C-trifluralin, ^{14}C-metazachlor, ^{14}C-metamitron, and ^{14}C-sulcotrione showed low values of DT_{50} in the soil, ranging from 2.5 to 25.4 days. The descending order of persistence was as follows: ^{14}C-metamitron > ^{14}C-trifluralin > ^{14}C-glyphosate > ^{14}C-metazachlor > ^{14}C-sulcotrione. It was observed that the degradation of ^{14}C-glyphosate, ^{14}C-trifluralin, and ^{14}C-sulcontrione depends on the high sorption strength of soil colloids. For ^{14}C-metamitron and ^{14}C-metazachlor, this relationship was not verified (Mamy et al. 2005).

The management adopted in the crop can also alter the persistence of an herbicide in the soil. Takeshita et al. (2019) evaluated the degradation of ^{14}C-aminocyclopyrachlor in an unamended and amended Alfisol-Paleudult with sugar cane straw. The addition of organic material in the soil contributed to increase the herbicide persistence. The DT_{50} of ^{14}C-aminocyclopyrachlor was prolonged with the highest content of organic carbon and the highest formation of bound ^{14}C-residues (non-extractable). The authors conclude by emphasizing the importance of further studies on herbicide degradation in the environment, especially in environments altered by soil management.

Herbicides are often applied in the mixture with other products. In this sense, it is importance quantify the degradation of herbicides that can be formed and mixed. Mendes et al. (2017) evaluated the fate of ^{14}C-mesotrione alone and mixed with S-metolachlor and terbuthylazine in two soils in Brazil. The degradation of the herbicide was faster and lower with the DT_{50}, without higher organic carbon and clay content, regardless of the application method. The sorption of ^{14}C-mesotrione is positively correlated with the organic carbon content. Therefore, it is possible that the ^{14}C-mesotrione available in the soil solution has been degraded and another portion of herbicide was sorbed in the colloidal fraction of the soil. On the other hand, the ^{14}C-mesotrione degradation alone or in mixture was similar in both soils.

^{14}C-diuron, ^{14}C-hexazinone, and ^{14}C-metribuzim were analyzed in five Brazilian soils. The lowest DT_{50} value for ^{14}C-diuron occurred in Nitosol

Eutrophic. In this soil, a greater cation exchange capacity (CEC) was observed; therefore, less degradation of the herbicide should be expected due to greater sorption of ^{14}C-diuron to soil colloids, since sorption of ^{14}C-diuron correlates positively with CEC. On the other hand, this higher CEC may favor growth and microbial activity due to the large amount of nutrients available, which would indirectly increase the degradation of ^{14}C-diuron (Guimarães et al. 2018).

^{14}C-hexazinone showed lower persistence in Oxisol Typic Hapludox, with a faster degradation rate than the other evaluated soils. The organic carbon content of this soil was lower than the others and possibly left ^{14}C-hexazinone more available in the soil solution at a level that increased accessibility to degrading microorganisms (Guimarães et al. 2018).

The degradation of ^{14}C-metribuzin did not correlate with soil attributes. However, the greater persistence and lower rate of degradation of this herbicide in the Typic Quartzipsaments suggest that the clay content influences the behavior of ^{14}C-metribuzin in the soil (Guimarães et al. 2018).

All the examples of evaluation of the degradation of ^{14}C-herbicides mentioned above show the complexity of the behavior of these products in the soil. Further details of the physical-chemical attributes of the soils, the rate of degradation (k), and the half-life time (DT_{50}) of the herbicides reported in these studies and in others are shown in Table 4.1.

4.11   Examples of radiorespirometry studies of microbial activity in soil

Respiration rate is an important parameter for evaluating biological activity in living systems (Kroukamp and Wolfaardt 2009). It is constantly taken into account when seeking to evaluate the biological characteristics of the soil, with the objective of implementing management practices. The effects of pesticides on soil microorganisms can also be evaluated by colony count, biomass, enzymatic activity, population changes, and microbial diversity (Imfeld et al. 2012, Jacobsen et al. 2014).

The most intensive agricultural practices depend on the use of herbicides to promote crop development, and a large proportion of these chemicals reach the soil. However, the continuous and extensive application of herbicides raises environmental concerns due to the effect on soil biological function (Cycón et al. 2011). Table 4.2 presents studies of the accumulated ^{14}C-CO_2 of ^{14}C-glucose microbial respiration of herbicide-treated soils worldwide.

The microbial degradation rate of herbicides in the soil is influenced by several factors; among these are the physical and chemical characteristics of the soil and the molecule itself, which act continuously determining

Table 4.1 Parameters of the degradation first-order kinetics (k and DT_{50}) of ^{14}C-herbicides in soils of the world

Local (country)	Soil classification	Soil Attributes			Radiolabeled herbicide	Specific activities (Bq mg^{-1})	C_0 (mg kg^{-1})	k (day^{-1})	DT_{50} (days)	Reference
		pH (H$_2$O)	OC (%)	Clay (%)						
France	Neoluvisol[a]	7.5	1.5	18.6	[^{14}C-phenyl]-acetochlor	1.25 × 10^6	n.a.$^{3/}$	0.140	5.1	Dictor et al. (2008)
	Calcosol[a]	8.0	1.6	23.9			n.a.	0.200	3.5	
Spain	Typic Haploxerept[b]	7.4	1.3	25.0	[ring-U-^{14}C]-prosulfocarb	3.16 × 10^6	4	0.086	8.0	Barba et al. (2019)
							10	0.142	13.9	
Brazil	Ultisol – Typic Hapludalf[b]	5.9	0.1	7.5	[U-phenyl-^{14}C]-quinclorac	1.21 × 10^4	0.2	0.012	57.8	Alonso et al. (2019)
	Entisol – Typic Haplowassents[b]	4.9	0.2	29.1				0.006	113.6	
	Oxisol – Typic Hapludox[b]	4.4	0.0	20.5				0.003	210.0	
	Entisol – Typic Quartzip[b]	4.5	0.0	2.5				0.003	266.6	
	Alfisol – Paleudult[b]	6.3	0.4	28.3				0.005	138.7	
United States	Lithic Xeric Haplocambids[b]	8.0	1.3	11.0	[phenyl-U-^{14}C]metsulfuron-methyl	1.43 × 10^6	0.2	0.015	15.0	Trabue et al. (2006)
	Typic Epiaquerts[b]	7.7	3.2	21.0				0.011	63.0	

(Continued)

Table 4.1 Parameters of the degradation first-order kinetics (k and DT_{50}) of ^{14}C–herbicides in soils of the world (*Continued*)

Local (country)	Soil classification	Soil Attributes			Radiolabeled herbicide	Specific activities (Bq mg^{-1})	C_0 (mg kg^{-1})	k (day^{-1})	DT_{50} (days)	Reference
		pH (H$_2$O)	OC (%)	Clay (%)						
France	Orthic Luvisol[a] *Depth (0–0.28 m)*	6.3	1.4	24.7	[ring-^{14}C]-isoproturon	1.28 × 10^7	0.1	n.a.	11.0	Alletto et al. (2006)
	Orthic Luvisol[a] *Depth (0.68– >1.20 m)*	8.6	0.5	19.3				n.a.	106.0	
	Calcil Cambisol[a] *Depth (0–0.28 m)*	6.5	1.4	27.8				n.a.	12.0	
	Calcil Cambisol[a] *Depth (0.63– >1.20 m)*	8.6	0.3	20.1				n.a.	203.0	
	Eutric Cambisol[a] *Depth (0–0.30 m)*	5.9	1.1	22.3				n.a.	10.0	
	Eutric Cambisol[a] *Depth (0.88– >1.20 m)*	8.4	0.4	33.9				n.a.	990	
France	n.a.	7.6	1.0	23.5	[methyl-^{14}C]-glyphosate	0.57	1.0	0.094	3.7	Mamy et al. (2005)
					[U-ring-^{14}C]-trifluralin	0.86		0.037	14.2	
					[U-phenyl-^{14}C]-metazachlor	0.83		0.174	3.5	
					[U-phenyl-^{14}C]-metamitron	0.81		0.018	25.4	
					[U-phenyl-^{14}C]-sulcotrione	0.68		0.203	2.5	

(*Continued*)

Table 4.1 Parameters of the degradation first-order kinetics (k and DT_{50}) of ^{14}C-herbicides in soils of the world (*Continued*)

Local (country)	Soil classification	Soil Attributes			Radiolabeled herbicide	Specific activities ($Bq\,mg^{-1}$)	C_0 ($mg\,kg^{-1}$)	k (day^{-1})	DT_{50} (days)	Reference
		pH (H_2O)	OC (%)	Clay (%)						
Brazil	Alfisol – Paleudult[b] (US[4/])	n.a.	1.7	45.8	[pyrimidine-2-^{14}C]-aminocyclopyrachlor	1.57×10^6	n.a.	0.004	187.3	Takeshita et al. (2019)
	Alfisol - Paleudult[b] (ASS[5/])	n.a.	4.0	45.8			n.a.	0.003	247.6	
Brazil	Alfisol – Paleudult[b]	6.4	1.8	37.6	[cyclohexane-2-^{14}C]-mesotrione	3.45×10^6	0.12	0.057	12.3	Mendes et al. (2017)
	Ultisol – Typic Hapludalf[b]	6.9	0.5	15.1				0.026	26.8	
	Alfisol – Paleudult[b]	6.4	1.8	37.6	[cyclohexane-2-^{14}C]-mesotrione+ S-metolachlor+ terbuthylazine		0.12+	0.053	12.9	
	Ultisol – Typic Hapludalf[b]	6.9	0.5	15.1			1.04+0.6	0.025	27.2	
Brazil	Oxisol Typic Hapludox[b]	5.1[10/]	1.8	72.9	[phenyl-U-^{14}C]-diuron	2.43×10^6	3.32	0.007	97.6	Guimarães et al. (2018)
	Oxisol Typic Hapludo[b]	4.5[10/]	1.0	75.4				0.005	128.4	
	Nitosol Eutrophic[b]	5.9[10/]	1.2	30.2				0.007	96.3	
	Udult soil[b]	5.1[10/]	1.6	32.7				0.007	103.5	
	Typic Quartzipsaments[b]	5.0[10/]	2.0	10.1				0.004	177.7	

(*Continued*)

Table 4.1 Parameters of the degradation first-order kinetics (k and DT_{50}) of ^{14}C–herbicides in soils of the world (*Continued*)

Local (country)	Soil classification	pH (H₂O)	OC (%)	Clay (%)	Radiolabeled herbicide	Specific activities (Bq mg⁻¹)	C_0 (mg kg⁻¹)	k (day⁻¹)	DT_{50} (days)	Reference
Brazil	Oxisol Typic Hapludox[b]	5.1[10/]	1.8	72.9	[ring-triazine-6-¹⁴C]-hexazinone	3.14 × 10⁶	0.41	0.008	83.5	Guimarães et al. (2018)
	Oxisol Typic Hapludox[b]	4.5[10/]	1.0	75.4				0.017	41.8	
	Nitosol Eutrophic[b]	5.9[10/]	1.2	30.2				0.011	66.0	
	Udult soil[b]	5.1[10/]	1.6	32.7				0.008	86.6	
	Typic Quartzipsaments[b]	5.0[10/]	2.0	10.1				0.007	100.5	
Brazil	Oxisol Typic Hapludox[b]	5.1[10/]	1.8	72.9	[ring-6-¹⁴C]-metribuzin	2.30 × 10⁶	1.2	0.009	81.5	Guimarães et al. (2018)
	Oxisol Typic Hapludox[b]	4.5[10/]	1.0	75.4				0.009	79.7	
	Nitosol Eutrophic[b]	5.9[10/]	1.2	30.2				0.009	75.3	
	Udult soil[b]	5.1[10/]	1.6	32.7				0.008	86.6	
	Typic Quartzipsaments[b]	5.0[10/]	2.0	10.1				0.005	138.6	

(*Continued*)

Table 4.1 Parameters of the degradation first-order kinetics (k and DT_{50}) of ^{14}C-herbicides in soils of the world (*Continued*)

Local (country)	Soil classification	Soil Attributes			Radiolabeled herbicide	Specific activities ($Bq\ mg^{-1}$)	C_0 ($mg\ kg^{-1}$)	k (day^{-1})	DT_{50} (days)	Reference
		pH (H_2O)	OC (%)	Clay (%)						
Brazil	Ultisol[b] (US[4/])	5.9	0.6	7.5	[P-methylene-^{14}C]-glyphosate	4.36×10^6	n.a.	0.018	37.9	Junqueira et al. (2019)
	Ultisol[b] (ASE[6/])	5.5	0.6	7.5				0.019	36.1	
	Alfisol[b] (US[4/])	6.3	4.4	28.3				0.018	37.9	
	Alfisol[2/] (ASE[6/])	6.1	4.1	28.3				0.032	21.3	
Brazil	Anthropogenic soil[b]	7.4	4.7	76.3	[phenyl-U-^{14}C]-diuron	2.43×10^6	0.02	0.010	66.7	Almeida et al. (2020)
	Ortic Quartzarenic[b]	6.7	0.3	2.0				0.008	88.9	
Spain	Typic Xerorthent[b] (US[4/])	7.7	0.5	n.a.	[ring-U-^{14}C]-linuron	9.62×10^6	2.0	0.019	49.8	Marín-Benito et al. (2014)
	Typic Xerorthent[b] (ASSS[7/])	7.3	2.0	n.a.				0.026	26.7	
	Typic Xerorthen[b/] (ASGM[8/])	7.4	2.7	n.a.				0.066	10.4	
	Typic Xerorthent[b] (ASSM[9/])	7.5	2.2	n.a.				0.039	30.9	

[a] According to Soil classification WRB/FAO
[b] According to American Soil Taxonomy

[3/]n.a., non-available; [4/]US, unamended soil; [5/]ASS, amended soil with sugarcane straw; [6/]ASE, amended soil eucalyptus-derived biochar; [7/]ASSS, amended soil with sewage sludge; [8/]ASGM, amended soil with grape marc; [9/]ASSM, amended soil with spent mushroom substrate; [10/]pH in $CaCl_2$.

Table 4.2 ^{14}C-CO_2 accumulated evolved from ^{14}C-glucose in microbial respiration of herbicide-treated soils worldwide

Local (country)	Soil classification	pH (H_2O)	OC (%)	Clay (%)	Herbicide	Dose herbicide (kg ha^{-1})	Incubation time (days)	Dose of ^{14}C-glucose (mg kg^{-1})	Specific activity (Bq mg^{-1})	Cumulative ^{14}C-CO_2 (% of added ^{14}C-glucose)	Reference
France	Eutric Calcic[a] Cambisol	7.8	1.1	33	DNOC (4,6-dinitroorthocresol)[d]	8	20	1	1.07×10^4	18.4	Rouard et al. (1996)
	Gleyic Luvisol[a]	7.4	0.1	19				10	9400	22.2	
								12	1.16×10^4	22.9	
China	n.a.	6.6	2.5	32	Bensulfuron-Methyl	10	60	1	5.68×10^7	59.6	Hou et al. (2009)
	n.a.	4.9	2.2	26						67.8	
Brazil	Oxisol Typic Hapludox[b]	5.0	1.8	72	Diuron	3.32	28	1	1.1×10^{10}	11.97	Dias (2012)
					Hexazinone	0.41				11.09	
					Metribuzin	1.2				10.14	
	Oxisol Typic Hapludox[b]	4.9	1.0	75	Diuron	3.32				11.25	
					Hexazinone	0.41				18.10	
					Metribuzin	1.2				11.03	
	Nitosol Eutrophic[b]	5.93	1.2	30.2	Diuron	3.32				11.1	
					Hexazinone	0.41				18.34	
					Metribuzin	1.2				9.67	
	Udult soil[b]	5.11	1.6	32.7	Diuron	3.32				11.07	
					Hexazinone	0.41				19.00	
					Metribuzin	1.2				11.07	
	Typic Quartzipsaments[b]	4.96	2.0	10.1	Diuron	3.32				12.87	
					Hexazinone	0.41				21.23	
					Metribuzin	1.2				12.85	

(Continued)

Table 4.2 ^{14}C-CO_2 accumulated evolved from ^{14}C-glucose in microbial respiration of herbicide-treated soils worldwide (*Continued*)

Local (country)	Soil classification	pH (H₂O)	OC (%)	Clay (%)	Herbicide	Dose herbicide (kg ha⁻¹)	Incubation time (days)	Dose of ¹⁴C-glucose (mg kg⁻¹)	Specific activity (Bq mg⁻¹)	Cumulative ¹⁴C-CO₂ (% of added ¹⁴C-glucose)	Reference
Brazil	Alfisol – Paleudult[c]	6.4	1.8	37.6	S-metolachlor	1.25	28	1	1.1×10^{10}	27.07	Mendes et al. (2018)
					terbuthylazine	0.75				26.15	
					Mesotrione	0.5				26.91	
					S-metolachlor + terbuthylazine+ mesotrione	1.25 + 0.75 + 0.5				27.46	
	Ultisol - Typic Hapludalf[c]	6.9	0.5	15.1	S-metolachlor	1.25				24.26	
					terbuthylazine	0.75				24.29	
					Mesotrione	0.5				23.34	
					S-metolachlor + terbuthylazine + mesotrione	1.25 + 0.75 + 05				25.30	
Brazil	Dark-red latosol[b]	6.7	3,6	63.9	Diuron	1.38	42	1	1.1×10^{10}	13.58	Reis et al. (2019)
					Hexazinone	0.39				11.36	
					sulfometuron-methyl	0.033				13.32	
	Typic Quartzipsaments[b]	6.8	1.3	62.0	Diuron	1.38				15.52	
					Hexazinone	0.39				11.96	
					sulfometuron-methyl	0.033				14.27	

n.a., non-available.

[a] According to soil classification WRB/FAO.
[b] According to Brazilian Soil Science Society.
[c] According to American Soil Taxonomy.
[d] A multi-use substance (herbicide, insecticide, acaricide, fungicide) used to control insect, pest and weed in agricultural crops.

its magnitude (Souza et al. 1999). Due to the variability of the physical, chemical, and biological characteristics of the soils, studies are based on the hypothesis that there would be a differential behavior in nature and the intensity of microbial activity in the presence of herbicides in different soils.

Mendes et al. (2018) observed that overall, the accumulated ^{14}C-CO_2 of microbial respiration for treatments in mixture S-metolachlor, ter-buthylazine, and mesotrione and alone was very similar in clay soil and sandy clay soil, showing that the addition of three herbicides in the mixture did not affect microbial herbicide respiration compared to treatment with each alone herbicide or control, without herbicides. However, Reis et al. (2019) observed that the application of diuron, hexazinone, and sul-fometuron-methyl alone or mixed in soils with different textures caused effects on soil microbial respiration. Hexazinone did not affect the evolution of ^{14}C-CO_2; however, diuron showed a higher evolution of ^{14}C-CO_2 in sandy and clayey soil, while sulfometuron-methyl led to an increase in respiration in sandy soil. The application of diuron, hexazinone and sulfometuron-methyl in mixture increased the respiration rate of sandy soil compared with the same treatment in clay soil or control, without her-bicide. Although glucose is a substrate of easy degradation by soil microorganisms, the research showed that in the presence of different herbicides, the respiration rate can be altered, especially in soils with sandy texture or with lower organic matter content.

The radiorespirometric assay for soils does not depend on direct measurements: the ^{14}C-CO_2 released from the sample is collected in a small volume of alkaline solution and quantified by liquid scintillation (Wang 1963, Alef 1995). However, some limitations are observed, for example, the difficulty of interpreting the origin of respiratory rate results, since only ^{14}C-CO_2 values evolved are found, and it is not possible to observe which metabolic processes occurred by increasing or decreasing the respiratory rate of soil microorganisms with the application of herbicides. Many microorganisms do not use amino acid isomers, which also makes it difficult to interpret soil microbiota radiorespirometry (Norris and Ribbons 1972).

The addition of a C source, in this case ^{14}C-glucose, can cause different results in the respiratory rate of soil microorganisms, which is an important point to be studied to integrate the structure of the soil microbial community with the different levels of glucose addition (Mendes et al. 2018). High glucose doses may decrease the amount of C incorporated by microbial biomass (Bremer and Kuikman 1994, Nguyen and Henry 2002) due to the impact on the energy status of the microbial community (Nguyen and Guckert 2001). However, at low concentrations, glucose is assimilated by microbial biomass but the energy and C supplied are insufficient for microbial growth (Bremer and Kuikman 1994, Hill et al. 2008). Schneckenberger et al. (2008) analyzed the amount of ^{14}C-glucose added to the soil and its effects on the accumulation of $^{14}C-CO_2$. For all ranges of easily available substrate concentrations that can occur in natural soils, the addition of increasing amounts of ^{14}C-glucose (0.0009, 0.26, 2.6, 26, and 257 µg Cg^{-1}) increased the mineralized portion in CO_2; however, the percentage of ^{14}C-glucose incorporated into microbial biomass generally decreased with the amount of added, especially in changes > 2.6 µg Cg^{-1}. Tian et al. (2015) reported that the high addition of glucose (204 µ Cg^{-1} in the soil) increased the percentage of ^{14}C-glucose mineralized in CO_2 but decreased the proportion of added glucose incorporated into microbial biomass. Thus, the added dose of ^{14}C-glucose is an essential factor to be established for analysis of radiorespirometry of a microbial community of the soil.

Some herbicide residues can serve as a source of carbon or energy for microorganisms that are degraded and assimilated by these microorganisms (Hussain et al. 2009, Blume and Reichert 2015). However, radiorespyrometric studies with ^{14}C-glucose do not allow determining what changes are taking place in the soil microbial community, since the respiratory rate of microorganisms may increase or decrease depending on the factors that contribute to such changes, including the degradation of compounds used as a source of nutrients, or xenobiotic poisoning, which lead to lower efficiency of carbon use, promoting, in both cases, an increase in CO_2 emissions (Santos et al. 2007).

Methodological adaptations should be studied to determine the respiratory rate of microorganisms in soils treated with herbicides. Pitombo et al. (2018) described a versatile methodology for measuring different gas flows (CH_4 and N_2O), in addition to CO_2, produced or consumed during breathing in the same sample, using gas chromatography. Its applications include the determination of the effects of different fertilizers on microbial respiration and the determination of the degradability of organic compounds, showing to be a promising alternative for evaluating the impacts of herbicides on soil microbiota.

4.12   Concluding remarks

Microorganisms are important in the degradation of herbicides, and the interaction between herbicides and microorganisms is influenced by soil attributes, climatic conditions, crop management, and herbicide dose. Each molecule has its particularity, being more or less bioavailable in the soil solution for degradation by microorganisms. The herbicide may be in the soil as bound residue and not be remobilized from the soil, or be biologically active, increasing its residual effect. In this sense, the role of microbial biomass played in the degradation of biologically active herbicides in the soil plays an important role in agricultural activities, determining the persistence of the molecule in the soil, which influences the weed control time, carryover, and environmental contamination risk.

Soil respirometric analysis using ^{14}C-glucose is a viable technique; however, few studies analyzing the mineralization of herbicides in the soil have been reported. The understanding of soil microbiological metabolism in the herbicide mineralization process needs to be better studied, since different factors can trigger a decrease or increase in respiratory rate, and it is important to analyze this process in detail.

Laboratory studies of herbicide degradation and microbial respiration in the soil are convenient and have predictive value, but real studies with field soil under natural conditions are essential to validate laboratory studies and confirm their results in the real condition. The use of ^{14}C labeled is basic for all herbicide degradation studies, mainly for registration of the product in the given country. However, risks and economics require studies with ^{14}C-herbicides to be carefully circumscribed.

References

Ahtiainen, J.H., P. Vanhala and A. Myllymäki. 2003. Effects of different plant protection programs on soil microbes. *Ecotoxicology and Environmental Safety* 54:56–64.

Alletto, L., Y. Coquet, P. Benoit and V. Bergheaud. 2006. Effects of temperature and water content on degradation of isoproturon in three soil profiles. *Chemosphere* 64:1053–1061.

Alef, K. 1995. Soil respiration. In: *Methods in Applied Soil Microbiology and Biochemistry*, eds. P. Alef and K. Nannipieri, 214–218. London: Academic Press.

Almeida, C.S., K.F. Mendes, L.V. Junqueira, F.G. Alonso, G.M. Chitolina and V.L. Tornisielo. 2020. Diuron sorption, desorption and degradation in anthropogenic soils compared to sandy soil. *Planta Daninha* 38:1–14.

Alonso, F. G., K.F. Mendes, L.V. Junqueira, V. Takeshita, C.S. Almeida and V.L. Tornisielo. 2019. Distribution and formation of degradation products of ^{14}C-quinclorac in five tropical soils. *Journal Archives of Agronomy and Soil Science* 1–12.

Anderson, J.P.E. 1975. Einfluss von temperatur und feuchte auf verdampfung, Abbau und festlegung von diallat im boden. *Zeitschrift fur Pflanznkrankheiten Pfanzenlschutz Sonderheft* 7:141–146.

Arias-Estévez, M., E. López-Periago, E. Martínez-Carballo, J. Simal-Gándara, J.C. Mejuto and L. García-Río. 2008. The mobility and degradation of pesticides in soils and the pollution of groundwater resources. *Agriculture, Ecosystems and Environment* 123(4):247–260.

Barba, V., J.M. Marín-Benito, C. García-Delgado, M.J. Sánchez-Martín and M.S. Rodríguez-Cruz. 2019. Assessment of ^{14}C-prosulfocarb dissipation mechanism in soil after amendment and its impact on the microbial community. *Ecotoxicology and Environmental Safety* 182:1–11.

Bartha, R. and D. Pramer. 1965. Features of a flask and method for measuring the persistence and biological effects of pesticides in soil. *Soil Science* 100(1):68–70.

Bending, G.D., S.D. Lincoln and R.N. Edmondson. 2006. Spatial variation in the degradation rate of the pesticides isoproturon, azoxystrobin and diflufenican in soil and its relationship with chemical and microbial properties. *Environmental Pollution* 139(2):279–287.

Blume, E. and J. M. Reichert. 2015. Banana leaf and glucose mineralization and soil organic matter in microhabitats of banana plantations under long-term pesticide use. *Environmental Toxicology and Chemistry* 34(6):1232–1238.

Bremer, E. and P. Kuikman. 1994. Microbial utilization of [14]C[U]glucose in soil is affected by the amount and timing of glucose additions. *Soil Biology and Biochemistry* 26(4):511–517.

Cycoń, M., A. Lewandowska and Z. Piotrowska-Seget. 2011. Comparison of mineralization dynamics of 2,4-dichlorophenoxyacetic acid (2,4-D) and 4-chloro-2-methylphenoxyacetic acid (MCPA) in soils of different textures. *Polish Journal of Environmental Studies* 20:293–301.

Dan, H.A., L.G.M. Dan, A.L.L. Barroso, S.O. Procópio, J. Oliveira, R.L. Assis and C. Feldkircher. 2011. Effect of the residual activity of pre-emergent herbicides applied in soybean on pearl millet cultivated in succession. *Planta Daninha* 29(2):437–445.

Dias, A.C.R. 2012. Leaching, mobility, degradation, mineralization and microbial activity of herbicides as a function of attributes of five soil types. Thesis (PhD in Crop Science), College of Agriculture "Luiz de Queiroz", University of São Paulo, Piracicaba, Brazil.

Dictor, M.C., N. Baran, A. Gautier and C. Mouvet. 2008. Acetochlor mineralization and fate of its two major metabolites in two soils under laboratory conditions. *Chemosphere* 71:663–670.

Food and Agriculture Organization of the United Nations (FAO). 2019. Pesticides Use. http://www.fao.org/faostat/en/#data/RP (accessed March 03, 2020).

Gebler, L. and C.A. Spadoto. 2008. Comportamento ambiental dos herbicidas. In: *Manual de Manejo e Controle de Plantas Daninhas* eds. L. Vargas and E.S. Roman, 39–69. Passo Fundo: Embrapa Trigo.

Guimarães, A.C.D., K.F. Mendes, F.C. Reis, T.F. Campion, P.J. Christoffoleti and V.L. Tornisielo. 2018. Role of soil physicochemical properties in quantifying the fate of diuron, hexazinone, and metribuzin. *Environmental Science and Pollution Research* 25:12419–12433.

Harvey Jr., J. 1983. A simple method of evaluating soil breakdown of ^{14}C-pesticides under field conditions. *Residue Reviews* 85:149–158.

Hill, P.W., J.F. Farrar and D.L. Jones. 2008. Decoupling of microbial glucose uptake and mineralization in soil. *Soil Biology and Biochemistry* 40(3):616–624.

Hou, X.W., J.J. Wu, J.M., Xu and C.X. Tang. 2009. Interactive effects of lead and bensulfuron-methyl on decomposition of ^{14}C-glucose in paddy soils. *Pedosphere* 19(5):577–587.

Hussain, S., T. Siddique, M. Saleem, M. Arshad and A. Khalid 2009. Impact of pesticides on soil microbial diversity, enzymes, and biochemical reactions. *Advances in Agronomy* 102:159–200.

IBAMA - Instituto Brasileiro do Meio Ambiente e dos Recursos Naturais Renováveis. Portaria IBAMA n° 6 de 17 de maio de 2012. DOU 23/05/2012. 2012. https://sogi8.sogi.com.br/Arquivo/Modulo113.MRID109/Registro58830/portaria%20ibama%20n%C2%BA%2006,%20de%2017-05-2012.pdf (accessed April 23, 2020).

Imfeld, G. and S. Vuilleumier. 2012. Measuring the effects of pesticides on bacterial communities in soil: a critical review. *European Journal of Soil Biology* 49:22–30.

Ismail, B.S., O.K. Eng and M.A. Tayeb. 2015. Degradation of triazine-2-^{14}C metsulfuron–methyl in soil from an oil palm plantation. *PLOS ONE* 10:e0138170.

Jacobsen, C.S. and M.H. Hjelmsø. 2014. Agricultural soils, pesticides and microbial diversity. *Current Opinion in Biotechnology* 27:15–20.

Jones, D.L. and A.C. Edwards. 1998. Influence of sorption on the biological utilization of two simple carbon substrates. *Soil Biology and Biochemistry* 30(14):1895–1902.

Junqueira, L.V., K.F. Mendes, R.N. Sousa, C.S. Almeida, F.G. Alonso and V.L. Tornisielo. 2019. Sorption–desorption isotherms and biodegradation of glyphosate in two tropical soils aged with eucalyptus biochar. *Archives of Agronomy and Soil Science* 1:1–17.

Khatoon, H., P. Solanki, M. Narayan, L. Tewari and J. Rai. 2017. Role of microbes in organic carbon decomposition and maintenance of soil ecosystem. *International Journal of Chemical Studies* 5(6):1648–1656.

Kroukamp, O. and G.M. Wolfaardt. 2009. CO_2 production as an indicator of biofilm metabolism. *Applied and Environmental Microbiology* 75(13): 4391–4397.

Laskowski, D.A., R.L. Swann, P.I. McCall and H.D. Bidlack. 1983. Soil degradation studies. *Residue Reviews* 85:139–148.

Lavorenti, A., A.A. Rocha, F. Prata, J.B. Regitano, V.L. Tornisielo and Pinto, O.B. 2003. Comportamento do diclosulam em amostras de um latossolo vermelho distroférrico sob plantio direto e convencional. *Revista Brasileira de Ciência do Solo* 27(1):183–190.

Law, S.E. 2001. Agricultural electrostatic spray application: a review of significant research and development during the 20th century. *Journal of Electrostatics* 51:25–42.

Luchini, L. and M.M. Andréa 2018. The use of nuclear techniques for environmental studies. In: *Integrated Analytical Approaches for Pesticide Management* eds. B. Maestroni and A. Cannavan, 165–182. Cambridge: Academic Press.

Mahía, J., A. Cabaneiro, T. Carballas and M. Díaz-Raviña. 2008. Microbial biomass and C mineralization in agricultural soils as affect by atrazine addition. *Biology and Fertility of Soils* 45(1):99–105.

Majewsky, M., T. Galle, L., Zwank and K. Fischer. 2010. Influence of microbial activity on polar xenobiotic degradation in activated sludge systems. *Water Science and Technology* 62(3):701–707.

Maier, R.M. 2000. Microorganisms and organic pollutants. In: *Environmental Microbiology* eds. R.M. Maier, I.L. Pepper, and C.P. Gerba, 363–402. San Diego: Academic Press.

Mamy, L., E. Barriuso and B. Gabrielle. 2005. Environmental fate of herbicides trifluralin, metazachlor, metamitron and sulcotrione compared with that of glyphosate, a substitute broad spectrum herbicide for different glyphosate-resistant crops. *Pest Management Science* 61:905–916.

Marín-Benito, J.M., E. Herrero-Hernández, M.S. Andrades, M. J. Sánchez-Martín and M.S. Rodríguez-Cruz. 2014. Effect of different organic amendments on the dissipation of linuron, diazinon and myclobutanil in an agricultural soil incubated for different time periods. *Science of the Total Environment* 476–477:611–621.

Mendes, K.F., B.A.B. Martins, M.R. Reis, R.F. Pimpinato and V.L. Tornisielo. 2017. Quantification of the fate of mesotrione applied alone or in a herbicide mixture in two Brazilian arable soils. *Environmental Science and Pollution Research* 24:8425–8435.

Mendes, K.F., S.A. Collegari, R.F. Pimpinato and V.L. Tornisielo. 2018. Glucose mineralization in soils of contrasting textures under application of S-metolachlor, terbuthylazine, and mesotrione, alone and in a mixture. *Bragantia* 77(1):152–159.

Mueller, T.C. and S.A. Senseman. 2015. Methods related to herbicide dissipation or degradation under field or laboratory conditions. *Weed Science* 63:133–139.

Norris, J.R. and D.W. Ribbons. 1972. *Methods in Microbiology*. London: Academic Press.

Nguyen, C. and F. Henry. 2002. A carbon-14-glucose assay to compare microbial activity between rhizosphere samples. *Biology and Fertility of Soils* 35(4):270–276.

Nguyen, C. and A. Guckert. 2001. Short-term utilisation of [14]C-[U]glucose by soil microorganisms in relation to carbon availability. *Soil Biology and Biochemistry* 33(1):53–60.

Organisation for Economic Co-Operation and Development (OECD). 2000. *Guidelines for Testing of Chemicals - Soil microorganism: carbon transformation test*. Paris: OECD. 10 p. (Test 217).

Organisation for Economic Co-Operation and Development (OECD). 2002a. *Guidelines for Testing of Chemicals - Aerobic and anaerobic transformation in soil*. Paris: OECD. 17 p. (Test 307).

Organisation for Economic Co-Operation and Development (OECD). 2002b. *Guidelines for Testing of Chemicals – Revised Proposal for a New Guideline*, 312. Paris: OECD. 15 p.

Oliveira Jr., R.S., W.C. Koskinen, C.D. Graff, J.L. Anderson, D.J. Mulla, E.A. Nater and D.G. Alonso. 2013. Acetochlor persistence in surface and subsurface soil samples. *Water Air Soil Pollution* 224(10):1–9.

Pitombo, L.M., J.C. Ramos, H.D. Quevedo, K.P. Carmo, J.M. Paiva, E.A. Pereira and J.B. Carmo. 2018. Methodology for soil respirometric assays: step by step and guidelines to measure fluxes of trace gases using microcosms. *MethodsX* 5:656–668.

Posen, P., A.A. Lovett, K.M. Hiscock, S. Evers, R. Ward and B.J. Reid. 2006. Incorporating variations in pesticide catabolic activity into a GIS-based groundwater risk assessment. *Science of theTotal Environment* 367:641–652.

Reis, M.R., A.A. Silva, M.D. Costa, A.A. Guimarães, E.A. Ferreira, J.B. Santos and P.R. Cecon. 2008. Atividade microbiana em solo cultivado com cana-de-açúcar após aplicação de herbicidas. *Planta Daninha* 26(2):323–331.

Reis, F.C., V.L. Tornisielo, B.A.B. Martins, A.J. Souza, P.A.M. Andrade, F.D. Andreote, R.F. Silveira and R. Victoria Filho. 2019. Respiration induced by substrate and bacteria diversity after application of diuron, hexazinone, and sulfometuron-methyl alone and in mixture. *Journal of Environmental Science and Health, Part B* 54(7):560–568.

Reid, B.J., C.J. MacLeod, P.H. Lee, A.W. Morriss, J.D. Stokes and K.T. Semple. 2001. A simple ^{14}C-respirometric method for assessing microbial catabolic potential and contaminant bioavailability. *FEMS Microbiology Letters* 196(2):141–146.

Rouard, N., M.C. Dictor, R. Chaussod and G. Soulas. 1996. Side-effects of herbicides on the size and activity of the soil microflora: DNOC as a test case. *European Journal of Soil Science* 47(4):557–566.

Saleem, M. and L.A. Moe. 2014. Multitrophic microbial interactions for eco- and agro-biotechnological processes: theory and practice. *Trends in Biotechnology* 32(10):529–537.

Santos, E.A., J.B. Santos, L.R. Ferreira, M.D. Costa and A.A. Silva. 2007. Phyto-stimulation by *Stizolobium aterrimum* as remediation of soil contaminated with trifloxysulfuron-sodium. *Planta Daninha* 25(2):259–265.

Souza, A.P., F.A. Ferreira, A.A. Silva, A.A. Cardoso and H.A. Ruiz.1999 Respiração microbiana do solo sob doses de glyphosate e de imazapyr. *Planta Daninha* 17:387–398.

Schneckenberger, K., D. Demin, K. Stahr and Y. Kuzyakov. 2008. Microbial utilization and mineralization of [^{14}C]glucose added in six orders of concentration to soil. *Soil Biology and Biochemistry* 40(8):1981–1988.

Sheehan, D. 1997. *Bioremediation Protocols*. Totowa: Humana Press.

Takeshita, V., K.F. Mendes, L.V. Junqueira, R.F. Pimpinato and V.L. Tornisielo. 2019. Quantification of the fate of aminocyclopyrachlor in soil amended with organic residues from a sugarcane system. *Sugar Tech* 1–9.

Trabue, S.L., D.E. Palmquist, T.M. Lydick and S.K. Singles. 2006. Effects of soil storage on the microbial community and degradation of metsulfuron-methyl. *Journal of Agricultural and Food Chemistry* 54:142–151.

Tian, J., J. Pausch, G. Yu, E. Blagodatskaya, Y. Gao and Y. Kuzyakov. 2015. Aggregate size and their disruption affect ^{14}C-labeled glucose mineralization and priming effect. *Applied Soil Ecology* 90:1–10.

United States Environmental Protection Agency (U.S. EPA). 2008a. *Guidelines for Fate, Transport and Transformation. Test - Terrestrial Field Dissipation*. United States: EPA. 50 p. (Series 835).

United States Environmental Protection Agency (U.S. EPA). 2008b. *Guidelines for Fate, Transport and Transformation. Test - Aerobic and Anaerobic Soil Metabolism*. United States: EPA. 19 p. (Series 835).

United States Environmental Protection Agency (U.S. EPA). 2020. Degradation Kinetics Equations. https://www.epa.gov/pesticide-science-and-assessing-pesticide-risks/degradation-kinetics-equations (accessed May 15, 2020).

Vivian, R., M.R. Reis, A. Jakelaitis, A.F. Silva, A.A. Guimarães, J.B. Santos and A.A. Silva. 2006. Persistência de sulfentrazone em Argissolo Vermelho-Amarelo cultivado com cana-de-açúcar. *Planta Daninha* 24(4):741–750.

Wang, W., Y. Wang, Z. Li, H. Wang, Z. Yu, L. Lu and Q. Ye. 2014. Studies on the anoxic dissipation and metabolism of pyribambenz propyl (ZJ0273) in soils using position-specific radiolabeling. *Science of the Total Environment* 472:582–589.

Wang, C.H. 1963. Metabolism studies by radiorespirometry. In: *Advances in Tracer Methodology* ed. S. Rothchild, 274–290. Springer: Boston.

Wolf, D.C., J.O. Legg and T.W. Boutton. 1994. Isotopic methods for the study of soil organic matter dynamics. *Methods of Soil Analysis: Part 2 Microbiological and Biochemical Properties* 5:865–906.

chapter five

Absorption, translocation, and metabolism studies of herbicides in weeds and crops

Ricardo Alcántara de la Cruz[1], Gabriel da Silva Amaral[1], Kassio Ferreira Mendes[2], Antonia María Rojano-Delgado[3], Rafael De Prado[3], Maria Fátima das Graças Fernandes da Silva[1]
[1]Department of Chemistry, Federal University of São Carlos
[2]Department of Agronomy, Federal University of Viçosa
[3]Department of Agricultural Chemistry and Edaphology, University of Cordoba

Contents

5.0   Introduction

Herbicides are a broad class of phytotoxic pesticides, usually chemicals with varying degrees of specificity, used to kill or inhibit the growth of weeds in both cultivated and non-cultivated areas (Gupta 2011; Duke and Dayan 2018). The 2,4-dichlorophenoxyacetic acid (2,4-D) was the first synthetic herbicide discovered and put into use at the end of the 1940s. In the next 20 years after the introduction of this herbicide, more than 100 new chemicals were synthesized, developed, and put into use (Gupta 2011). From 1950 to 1980, a new herbicidal mechanism of action (MOA) was discovered, on average, every 2 or 3 years (Dayan 2019). The easy use of these products and cost savings in labor compared to traditional weed management methods, as well as the increase in their acceptance and availability, made herbicides the main weed management tool (Duke 2012). Today, herbicides are one of the most important inputs in agricultural production, so much so that they occupy 60% of the global pesticide market, and weed management is handled almost exclusively with these products in most intensive and large-scale crop production systems (Dayan 2019). However, no new MOAs have been marketed in the last three decades (Duke 2012).

Herbicides caused great changes in weed management, and they have contributed to increasing the world production of food, fibers, and fuels in efficient, economic, and environmentally sustainable ways (Nandula and Vencill 2015). One of the main principles in designing herbicide formulations is to carry the active ingredient through the barrier of plant organs to reach its target site by short- or long-distance translocation (Devine et al. 1993; Nandula and Vencill 2015). Herbicides are formulated to control weeds, and their efficacy depends upon the dosage (Singh et al. 2020), but at the same time they must be safe for crops, with minimal risk to human health, fauna, flora, and the environment (Nandula and Vencill 2015). Before placing a herbicide on the market, these products are subjected to rigorous testing of their toxicological, residual, physicochemical, and biological properties to comply with a series of legal provisions (Nandula and Vencill 2015), depending on the country, to achieve a correct, effective, and efficient use of these products in protecting crops. The absorption and translocation of herbicides are phenomena widely studied in crops by employing analytical techniques and complementary approaches to determine the location of the product in plant tissues and organelles (Nandula and Vencill 2015).

Changes in herbicide absorption and translocation patterns, as well as in their metabolism, are biochemical and physiological events reported as resistance mechanisms to different herbicides in weeds (Mendes et al. 2017b). According to the Weed Science Society of America (WSSA), herbicide resistance is the inherited ability of a plant to survive and reproduce following exposure to a dose of herbicide that is normally lethal to wild-type plants of the same weed species (WSSA 1998). Such resistance can be conferred by resistance mechanisms of the target site resistance (TSR), non-target site resistance (NTSR), or by the two types of mechanisms. Most herbicide target sites are enzymes found within plant cells (Moss 2017). The TSR are molecular events (nucleotide polymorphisms in the herbicide target gene and/or increased gene copy number that induces target enzyme overexpression) limiting the binding or interaction of the herbicide with the target enzyme or requiring a greater herbicide concentration to inhibit it (Gaines et al. 2019, 2020). Reduced absorption, impaired translocation, and enhanced metabolism of herbicides are resistance mechanisms of the NTSR type that occur before the herbicide reaches its target site. NTSR mechanisms have gained great relevance in recent years, because herbicide resistance cases involving these mechanisms are becoming more frequent around the world, and the knowledge about the multigenic complexes that regulate them is still limited (Ghanizadeh and Harrington 2017; Gaines et al. 2020). Therefore, the evaluation of these biochemical and physiological events is an important component in the study of herbicide resistance (Nandula and Vencill 2015).

Radioisotopes are used as tracers in various ways in agriculture, engineering, medicine, etc., to determine biological pathways and mass balance studies of organic compounds. The most widely used radioisotopes in plant sciences are tritium (^{3}H) and ^{14}C, which can be precisely differentiated by their radiation signal without any noise contribution from unlabeled compounds, even when both show similar chemical behavior (Freud and Hegeman 2017).

With the availability of herbicides radiolabeled with ^{14}C since the 1950s, autoradiographic and counting methods have simplified and expanded research approaches on herbicide absorption, translocation, and metabolism in crop and weed plants (Yamaguchi and Crafts 1958; Nandula and Vencill 2015), as well as to assess the fate of these molecules in the environment (Mendes et al. 2017a). Understanding these plant biochemical and physiological events is essential to plan weed management strategies as well as improve product effectiveness. Therefore, the objective of this chapter is to integrate the knowledge about the study of the absorption, translocation, and metabolism of herbicides in plants by describing the principal theoretical concepts and the research procedures related to the use of ^{14}C-herbicides.

5.1 Concepts

Radioisotope: Atoms of an element with the same atomic number but with different masses and with excess nuclear energy. Radioisotopes have the same chemical properties (same number of protons) as

stable atoms of the same element but not retain its physical properties (different number of electrons). Nuclei of radioisotopes emit
energy in the form of ionizing radiation as they seek a more stable
configuration (Freud and Hegeman 2017).

Specific activity: Inherent physical property of a certain radioisotope;
defined as the radioactivity per unit mass or quantity of matter
(mol), indicating the abundance of a radioisotope with respect to
the total number of atoms of that element (de Goeij and Bonardi
2005). The specific activity of any compound is expressed in dpm/
mol as specific activity $= ln2/t_{1/2} \times 6.0221 \times 10^{23} \times n$, where $t_{1/2}$ is the
half-life of the radioactive isotope (min), 6.0221 is Avogadro's num-
ber, and n is the average number of isotopes per molecule (Penner
et al. 2009).

Mass balance: Balance between the amount of radiolabeled product
applied to a known specific activity and the amount recovered,
which is expected to be the total applied to achieve said balance;
i.e., if 100 radioactive units are applied, it is expected to recover the
100 units. However, radiolabeled product may be lost due to volatilization, radioactive dust, root exudation, and/or by handling during
experimental techniques. Working with radiolabeled herbicides, the
mass balance is expressed as a percentage of recovery (sum of the
radioactivity measured in all parts of the plant, leaf washes [or
roots], and root exudates) in relation to the radioactivity applied at
the beginning of the experiment, and it is expected to be greater
than 80% (Nandula and Vencill 2015).

Herbicide absorption: Also referred to as herbicide penetration or
herbicide uptake, is the amount or percentage of the active ingredient that has entered into a treated plant from that amount of
herbicide that was retained on the surfaces of aerial or under-
ground structures. Absorption rate is calculated as the sum of
the radioactivity measured in all parts of plants and is generally
expressed as percentage of applied radioactivity (Nandula and
Vencill 2015).

Herbicide translocation: Movement or transport of the active ingredient
from the point where it was absorbed toward other parts of the plant
using the phloem (simplast – inside the cells) and xylem (apoplast –
cell walls) systems (Menendez et al. 2014). Translocation rate is the
sum of radioactivity recovered in all plant parts except the treated
leaf (or roots) and is expressed as percentage of absorbed radioactivity (Nandula and Vencill 2015).

Herbicide metabolism: Transformation of herbicides by biochemical reactions, which are regulated by a large number of genes that encode
enzymes, resulting in the detoxification or bioactivation (secondary

metabolites) of parent molecules, neutralizing and/or reducing the harmful effects of herbicides on plants (Gaines et al. 2020).

Radiolabeled herbicide distribution: Radioactivity accumulated in a predetermined part of the treated plant including treated leaf (or root). Distribution is recommended to be expressed as a percentage of absorbed radioactivity (Nandula and Vencill 2015).

Combustion: Incineration of solid samples (plant tissue) containing radioactive material (radiolabeled herbicide) in an automatic preparation and oxidation system or biological oxidizer. This equipment creates an atmosphere rich in hydrogen and oxygen, supplied externally, in its combustion chamber where the carbon, labeled or not with ^{14}C, contained in the plant sample is oxidized into carbon dioxide (Mendes et al. 2017b). The combustion chamber is airtight, allowing $^{14}CO_2$ released from incineration to be trapped in a radioactive dioxide absorber solution.

Phosphor storage (PS) screen: Reusable plates with a durable cellulose acetate coating and a photostimulable phosphor layer that trap and store energy from ionizing radiation from radioisotopes of solid samples in a stable state. Radiation causes excitation of phosphor luminescent ions, generating a latent image (Barthe et al. 2012).

Autoradiography: Digitalization of the latent image from the excited phosphor luminescent ions by ionizing radiation on the reusable PS-screen. The excited phosphor luminescent ions emit light proportionally to the absorbed radiation from each location on the PS-screen, which is detected/scanned by a single confocal optic with a laser in a PS-system whose wavelength is in its absorption range, to create a high-resolution digital image (file) with quantitative data (Perkin-Elmer 2020).

Cold herbicide treatment: Herbicide application (without radioisotope) on plants prior to treatment with ^{14}C-herbicide. The non-labeled herbicide must be a trade formulation and is sprayed on plants following the manufacturer's recommendations for field applications, i.e., dose, volume, pH adjustment of water, and use of an adjuvant(s) if indicated in the label. Sublethal doses such as the ED_{50} (effective mean dose that reduces the dry/fresh weight by 50%) may be used when a weed population is extremely sensitive to the full herbicide dose (Nandula and Vencill 2015).

Hot herbicide treatment: Application of the ^{14}C-herbicide solution of the leaves that were predetermined to receive the treatment. The herbicide solution is prepared by mixing ^{14}C-herbicide with a trade herbicide formulation, i.e., ^{14}C-herbicide + trade herbicide. The final concentration of herbicide (g ia ha^{-1}) must be the same as that applied in cold treatment, and the minimum specific activity must be ≤170 Bq plant⁻¹ (10,000 dpm) (Nandula and Vencill 2015; Mendes et al. 2017b).

5.2   Herbicide route in the plant

Herbicides reach their target (weeds) directly or indirectly. Post-emergent herbicides are sprayed directly on their target, i.e., foliage and stems of weeds. Pre-emergent herbicides first reach the ground before coming into contact with the plumules, radicles, rhizomes, roots, seeds, stolons, or tubers (Mendes et al. 2017b). When the herbicides reach the plant, they can be absorbed via leaf and root, depending on the method of application and action mode of herbicide, and then herbicides exert their phytotoxic effect near the point of entry or are translocated throughout the plant (Menendez et al. 2014). Understanding the foliar retention rates and penetration process of herbicide has allowed the formulations to be optimized by improving the transport of the active ingredient through the cuticle to the apoplast, which subsequently reaches the symplast, where the phloem transfers it to the rest of the plant (Hess 2018). Figure 5.1 summarizes the consecutive steps of the route of a herbicide from the application until it reaches its target site within the plant cell, as well as the physiological,

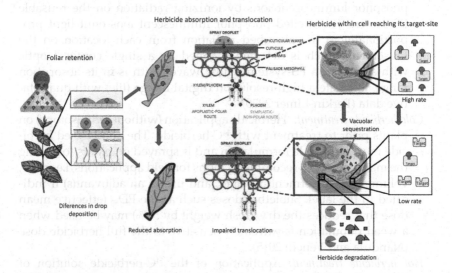

Figure 5.1 Herbicide route from the application up to reach its target site (top), and physiological, biochemical, and metabiological events, which can occur in isolation or together, that prevent/limit the herbicide from reaching its target site at lethal rates (bottom), referred to as non-target-site resistance mechanisms (NTSR). (Adapted from Délye 2013.)

biochemical, and metabolomical events that can occur before the herbicide reaches its target site, which are described below.

5.3 Foliar retention

Foliar retention is the capture of droplets of herbicide solution from application by leaves, which determines the amount of active ingredient in a plant (Yao et al. 2014). Foliar retention is determined by external factors (temperature, light, relative humidity, and soil moisture), physicochemical properties of the herbicide formulation, as well as inherent characteristics of plant surface (density and distribution of trichomes, leaf roughness, stoma geometry, and wax composition and deposition) (Alcántara-de la Cruz et al. 2016; Palma-Bautista et al. 2020). The leaf cuticle is a thin lipidic layer composed by insoluble cutin and soluble waxes, which covers the aerial parts of the plants to avoid the loss of water by evapotranspiration (Kosma et al. 2009). The epicuticular waxes conforming the outer surface of the cuticle are hydrophobic (Menendez et al. 2014); therefore, the combination of cuticular hydrophobicity, leaf surface characteristics, and water tension determine the amount of herbicide solution retained on the leaves (Burkhardt et al. 2012).

Differences in the composition of the cuticle as well as in the formulation may result in different susceptibility to herbicides among and within plant species (Hess and Falk 1990; Menendez et al. 2014). For instance, foliar retention of glufosinate in *Ambrosia artemisiifolia* was similar to the retention rate of water alone (400 μL g^{-1} dry matter), but glyphosate retention was almost double due to the surfactants present in the evaluated commercial formulation (Grangeot et al. 2006); the addition of the ethoxylated amine surfactant increased by 25% to 30% the foliar retention of glyphosate in barley, proportionally improving the efficacy of the herbicide (Gauvrit 2003): the effectiveness (~42%), foliar retention (~45%), and absorption (~12%) of glyphosate increased in *Lolium rigidum* and *Conyza canadensis* by adding an adjuvant based on nethoxyl alcohols polyglycol and aryl polyethoxyethanol (Palma-Bautista et al. 2020). In addition, foliar retention may be dependent on the phenology of the plants, as observed in the *Conyza* species, where foliar retention was greater during the elongation of the stem than during flowering (González-Torralva et al. 2010).

Pesticide retention rate on the leaves is an important trait to ensure the effectiveness of these products (Yao et al. 2014). Due to the inherent features of the leaf surfaces of each species, some plants are considered easier to moisten than other ones (Michitte et al. 2007; Yao et al. 2014); therefore, differences in foliar retention between species, whether weed or cultivated, are common. For example, *Avena fatua* retained fourfold more herbicide (asulam) than *Linum usitatissimum* (Sharma et al. 1978); *Amaranthus hybridus* retained sixfold more imazaquin than *Senna obtusifolia* when adjuvant were not added (Reddy and Locke 1996); *Eleusine indica* retained 40% to 65% more quinclorac than *Digitaris sanguinalis* (Zawierucha and Penner 2000), among other cases. For this reason, the recommended field doses of herbicides may vary depending on the crop system and the weed species to be controlled.

On the other hand, differences in foliar retention between herbicide susceptible and resistant weed populations of the same plant species have also been observed. Glyphosate-resistant populations of *L. multiflorum* and *L. perenne* retained between 17% to 45% less herbicide solution in relation to their respective susceptible counterparts (Michitte et al. 2007; Fernández et al. 2017); tribenuron-methyl resistant *Sinapis alba* retained 24% less herbicide solution (Rosario et al. 2011), among other species. These differences were possibly due to detrimental morphophysiological changes of resistant plants compared to susceptible ones, i.e., a fitness penalty in plant size, root anatomy, number of leaves and tillers, height, or flowering time, among other traits (Vila-Aiub 2019), and even possible changes in the production and deposition of cuticle waxes.

Foliar retention is a parameter little considered in herbicide resistance studies, because in most cases no differences between resistant and susceptible plants were observed, or sometimes resistant populations presented greater foliar retention, leading researchers to conclude that there is no relationship between foliar retention rate and herbicide resistance (Feng et al. 2004; González-Torralva et al. 2012; Jiménez et al. 2015). However, it is important to note that the fitness penalty associated with herbicide resistance is not universal, and its expression or lack thereof is dependent on the mechanism (NTSR or TSR) that governs such resistance, genetic background of the resistant population, and dominance of the fitness cost traits as well as environmental conditions (Vila-Aiub 2019); therefore, the evaluation of foliar retention of the herbicide spray solution must not be ignored in herbicide resistance studies.

5.4 Herbicide absorption

Prior to reaching its target site and causing a phytotoxic action, the herbicide must be absorbed by the plant (Sterling 1994). Foliar absorption differs from root absorption, since it implies that the herbicide must be absorbed by organs that are not specifically designed to absorb substances (Menendez et al. 2014). To be absorbed, the active ingredient of herbicide must migrate through multiple barriers of the leaves and roots, such as cuticle, epicuticular waxes, stomata and/or the external suberized (exodermis) and internal (Casparian strip) layers of the root, until they reach the apoplast and finally penetrate the plant cells (Hull, 1970; Devine et al. 1993; Menendez et al. 2014). The ability of a herbicide to penetrate the cuticle depends on the acidity, lipophilicity, mobility, and solubility of the molecule, as well as plant cell membranes and the electrochemical potential in the plant cell (Sterling 1994; Schreiber 2006); therefore, not all herbicide retained on plant surfaces is absorbed.

Herbicides can be absorbed by passive diffusion or by active transport. Most lipophilic, neutral molecule, or lipophilic herbicides move across lipidic membranes of the plant by passive diffusion (downward solute movement on an electrochemical gradient) (Sterling 1994). Passive diffusion of herbicides into plant tissues and cells consist in reach equilibrium concentrations. This equilibrium is reached rapidly because cell membranes are not an appreciable barrier to the herbicide, and these molecules have the ability to diffuse out of cells in solutions that do not contain the herbicide (Hess, 2018). Active transport is continuous over time, allowing the accumulation of herbicide molecules against a saturable concentration gradient as external herbicide concentrations increase, and requires energy from metabolic processes to move the herbicide molecule across the plant cell membrane against its electrochemical potential (Sterling 1994). This herbicide absorption process is limited to dalapon, 2,4-D, glyphosate, and paraquat (Hart et al. 1992), and usually involves a protein transporter located in the plant membrane, which recognizes herbicide as structures similar to the endogenous molecules carried by that transporter, carrying the herbicide into the plant cell. In addition, the saturable component of active transport is added to the passive diffusion (Sterling 1994).

Similar to foliar retention, herbicide absorption depends on external and internal factors and the physicochemical properties of the herbicide. External factors are not an inheritable source of herbicide resistance, while internal parameters are (Menendez et al. 2014), since any changes in the chemical composition of the cuticle can reduce the absorption patterns, decreasing the herbicide efficacy (Délye 2013). Reduced absorption, due to reduced retention or penetration of the herbicide through the cuticle, has been characterized as an NTSR mechanism conferring low herbicide resistance levels in different weeds (Moss 2017).

5.5   Herbicide translocation

After absorption by the root or shoot tissues, herbicide must reach its target site in a concentration sufficient to be lethal. To be translocated, the herbicide must move out of the treated leaf or root being necessary to enter plant tissue (Shaner 2009). The movement of herbicides in the cell walls can be by diffusion or with any flow of water mass. Herbicide movement between cells occurs passively through cell-to-cell cytoplasmic connections (plasmodesms) (Hess, 2018), following the direction of solute flow along with many other solutes in the phloem (Devine et al. 1993).

Herbicide translocation can be short- or long distance. Herbicide absorption can be considered short-distance translocation, because herbicides only need to penetrate a few cell layers to reach their target site. This translocation pattern is characteristic of contact herbicides such as pre-emergent growth inhibitors or post-emergent photosynthesis inhibitors. Movement of herbicide from the root surface to the xylem is an important short-distance translocation system (Hess, 2018). Long-distance translocation is characteristic of systemic herbicides such as glyphosate, acetolactate synthase inhibitors, and graminicides (Nandula and Vencill 2015), and enables the herbicide to reach nontreated parts of the plant. Long-distance translocation in the xylem and phloem becomes important for post-emergent herbicides used to control perennial or annual weeds beyond the seedling stage, because without an adequate translocation, the whole plant is not killed, allowing the weed to regrow (Hess, 2018).

Impaired translocation of a herbicide decreases the concentration that reaches the target site (Rojano-Delgado et al. 2014), significantly reducing the herbicide efficacy. Reduced translocation involves a restriction on the movement of the herbicide within the plant, often as a consequence of its compartmentation or sequestration into the vacuoles or plastids (Moss, 2017). Reduced herbicide translocation due to vacuolar sequestration is recognized as a major NTSR mechanism of resistance to glyphosate and paraquat (Ghanizadeh and Harrington 2017). Shaner (2009) proposed four potential mechanisms to explain the reduced cellular absorption and translocation of glyphosate: (1) a putative phosphate transporter that in nonresistant plants is responsible for the uptake of glyphosate, and loses affinity for the herbicide molecule, reducing glyphosate absorption and translocation; (2) evolution of a new transporter responsible for carrying glyphosate from the absorption point and sequestering it in the nearest vacuoles before glyphosate reaches its target site; (3) evolution of a trans- porter that expels glyphosate out of the cell and into the apoplast; and (4) evolution of a chloroplastic transporter that pumps glyphosate out of the chloroplast. For the vacuolar sequestration of paraquat, no spe- cific transporter involved has been identified, but evidence suggests the participation of transporters of the ATP-binding cassette (ABC), cationic amino acid (CAT) and L-type amino acid (LAT) (Fujita et al. 2012; Moretti et al. 2017). ABC-transporters have also been associated with the vacuolar sequestration of glyphosate (Tani et al. 2015), and although these genes do not seem to be directly involved in the resistance mechanism (Moretti et al. 2017), in both cases, rapid vacuolar sequestration of paraquat and/or glyphosate results in restriction of movement of the herbicides within the plant, preventing them from reaching their target sites.

5.6   Herbicide metabolism

Metabolism or biotransformation of herbicides is a complex multi-step process that involves the coordinated action of various types of enzymes, considered to be the main factor of selectivity and resistance to herbicides (Ghanizadeh and Harrington 2017; Gaines et al. 2020). Among the most important enzyme systems involved in enhanced herbicide metabolism are endogenous cytochrome P450 monooxygenases (Cyt-P450 or CYP), glutathione S-transferases (GST) and glucosyl transferases (GT) and/or other enzyme systems such as aryl acylamidase (Yu and Powles 2014; Nandula et al. 2019) in both cultivated and weed plants. In all cases, the genetic control of these enzymes is complex and is often regulated by multiple genes, i.e., herbicide metabolism is a polygenic NTSR mechanism. The clearest example is found in studies on the inheritance of resistance mediated by Cyt-P450 that have frequently identified the expression of hundreds of different isoenzymes in plants (Moss 2017). In addition, a single enzyme subfamily may be metabolizing herbicides of different chemical classes. For instance, CYP81A isoenzymes were able of metabolize 18 herbicides of acetolactate synthase (ALS), acetyl coenzyme A carboxylase (ACCase), phytoene desaturase (PDS), photosystem II (PSII), protoporphyrinogen oxidase (PPO), 4-hydroxyphenylpyruvate dioxygenase (HPPD), and 1-deoxy-D-xylulose-5-phosphate synthase (DXS) inhibitors by demethylation or hydroxylation reactions, showing the potential of Cyt-P450 enzymes to confer broad cross-resistance to herbicides (Dimaano et al. 2020).

Herbicide metabolism generally occurs in three phases. In phase I (primary metabolism), herbicide molecules are bioactivated in the endoplasmic reticulum by Cyt-P450s in less active compounds (detoxification), but occasionally they may be more phytotoxic metabolites (bioactivation). Phase II usually occurs in the cytoplasm, and the products of phase I are converted, mainly by GSTs and GTs, into even less toxic water-soluble conjugates by glucosyl, glutathione, or amino acid by conjugation with biomolecules such as glutathione/sugar. Metabolism of some herbicides ends in this phase. In phase III, conjugates of previous phases are transformed by further conjugation, breakup, or oxidation reactions into nontoxic secondary conjugates or insoluble bound residues, and they are subsequently transported by ABC other transporters to the vacuoles or cell walls, where additional breakdown or sequestration occurs (Nandula et al. 2019; Gaines et al. 2020).

Enhanced herbicide metabolism is one of the most important NTSR mechanisms and is the most studied resistance mechanism, especially among grass species (Délye 2013, Ghanizadeh and Harrington 2017). Both susceptible and resistant weed plants are capable of metabolizing herbicides; however, susceptible ones do so to a limited degree, while resistant plants survive because they have a greater and enhanced ability to metabolize them. These differences are generally determined quantitatively, which can make detection or interpretation of this type of resistance problematic, especially when enhanced herbicide metabolism occurs with other resistance mechanisms (Moss 2017).

5.7   Assessment of absorption, translocation, and metabolism in plants with radioisotopes

Since the first radiolabeled herbicides were synthesized in the 1950s, the majority of techniques used for absorption and translocation studies have been based on the uses of these compounds by counting and/or auto-radiographed methods to detect the cross- and longitudinal-sectional or general concentration and distribution patterns of herbicides in plant tissues (Yamaguchi and Crafts 1958; Sterling 1993; Mendes et al. 2017b). Most herbicides used in these studies are labeled with [14]C. Historically, ^{14}C-herbicide absorption and translocation studies in plants were implemented to evaluate the behavior of a new herbicide in a given plant species, comparing two or more herbicides, specific formulations, additives, or the effect of environmental conditions. However, due to the dramatic increase in herbicide-resistant weeds, these types of studies were adapted to evaluate these phenomena as resistance mechanisms (Nandula and Vencill 2015; Mendes et al. 2017b).

Absorption and translocation studies are usually carried out in greenhouses or growth chambers to reduce variability and improve the reproducibility of the experiments. If the experiments are conducted in a greenhouse, it is recommended that they be carried out when the outside temperatures are not prohibitively high and light intensity is moderate (temperate conditions), or if the plants are grown in a greenhouse and then taken to a growth chamber, plants must acclimatize to chamber conditions for at least 24 h before herbicide applications. In addition, repetition of experiments should be carried out under as similar conditions as possible (Nandula and Vencill 2015).

The most widely used quantitative method to evaluate the absorption and translocation of herbicides in plants involves the liquid scintillation spectrometry (LSS) technique (Figure 5.2a). This method determines the β radiation emitted by ^{14}C atoms when their molecules are excited after collision with high-energy particles. The energy of the decaying β particles is transformed into photons by the scintillator and detected and measured on the spectrometer by a photosensitive detector (photocathode) giving a signal proportional to the number of ^{14}C atoms. The commonly accepted unit of measurement for radioactivity is the becquerel (Bq = activity of a quantity of radioactive material in which one nucleus decays per second) (Nandula and Vencill 2015). However, most scintillation counters determine ^{14}C radioactivity in disintegration per min (dpm); therefore, it is important to consider the following conversions:

$$1 \text{ Bq} = 1 \text{ disintegration per second (dps)} = 60 \text{ dpm}$$

$$1 \text{ kBq} = 59,946 \ (5.946 \times 10^4) \text{ dpm}$$

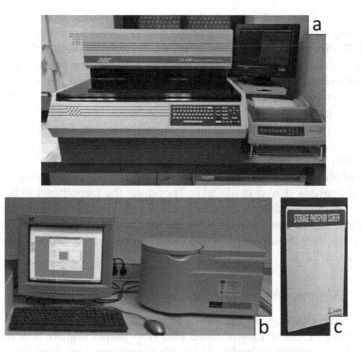

Figure 5.2 (a) Liquid scintillation counter (Beckman LS-6500, Beckman Coulter Inc.); (b) storage phosphor system (Cyclone Plus, Perkin-Elmer); and (c) reusable-storage phosphor screen from the Laboratory of "Herbicidas en el Ambiente" of the Department of Agricultural Chemistry and Edaphology, UCO/Spain.

Quantitative translocation analyses of ^{14}C-herbicides employ the phosphorimaging technique using a phosphor storage (PS) system that allow visualization of the displacement and location of the ^{14}C-herbicide within the plant (Figure 5.2b). This technique is a version of solid-state liquid scintillation making it possible to capture the energy of a radioactive solid matrix (Van Kirk et al. 2010). The principle of phosphorimaging is similar to traditional autoradiographic techniques, i.e., it relies on high-energy particle decay, but this technique is between 10 and 250 times more sensitive than x-ray film, depending on the radioactive isotope used (Van Kirk et al. 2010). The PS system performs digital images (files that can be printed, exported, and archived) of solid matrixes by laser-scanning reusable PS screens (Figure 5.2c) (Perkin-Elmer 2020). These screens capture and store ionizing radiation from ^{14}C-herbicides that is found within the tissues of treated plants. The PS system is relatively expensive; however, it is faster and safer compared to traditional autoradiographic techniques, since it does not require manipulation of chemicals that are harmful to health (Mendes et al. 2017b). Although phosphorimaging autoradiographies of ^{14}C-herbicide treated plants are used mainly as a qualitative result that accompanies and supports the quantitative results obtained from LSS analyses, quantitative analysis of the translocation may also be done in the software provided together with the PS system (Perkin-Elmer 2020). The translocation is expressed as the rate between the percentage of signal intensity on the applied zone as well as above and below it, and the total signal intensity on a defined image containing ^{14}C-herbicide (Mendes et al. 2017b).

Both qualitative and quantitative methods are destructive experiments; i.e., treated plants are cut/sectioned at different intervals after treatment in order to determine the absorption and translocation patterns of the herbicide in the plant considering proper planning and statistical analysis (Kniss et al. 2011). Therefore, the hypothesis to be tested and experimental design must be carefully defined before beginning any experiment; because radioactive herbicides are expensive and difficult to eliminate, they should only be used for clear research purposes (Nandula and Vencill 2015).

5.7.1   Absorption and translocation

The outline practices and procedures to study the absorption and translocation of ^{14}C-herbicides in weed and crop plants (Figure 5.3), resistant and/or susceptible to herbicides, with complementary highlights from Nandula and Vencill (2015) and Mendes et al. (2017b), are described in the text that follows.

Planning of experiments. Determine the intervals of evaluation, duration of experiments, and how many plants (repetitions) will be evaluated per plant population (resistant and susceptible) at each evaluation time. It is recommended to evaluate at least six harvest times in addition to the control (0 hour after treatment [HAT]), but under conditions of limited resources it is better to increase the number of evaluation times and reduce the number of repetitions ($n \geq 2$). Intervals between evaluation times depend on the type of herbicide to be evaluated. For example, imazamox and glyphosate are two systemic herbicides, but the first is absorbed and translocated faster than the second; therefore, evaluation times are usually different. Based on the literature, the highest rate of imazamox absorption occurs between 3 and 12 HAT, and that of translocation between 24 and 48 HAT (Domínguez-Mendez et al. 2019; Rojano-Delgado et al. 2019), while for glyphosate, these rates occur between 48 and 96 HAT.

Preparation of the plants before treatment of ^{14}C-herbicide. Choose homogeneous plants, both in size and in phenological stage, for treatment with ^{14}C-herbicide. The age and size of treated plants must be as

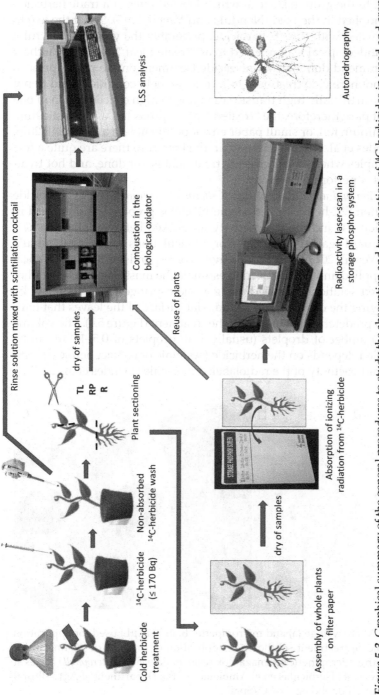

Figure 5.3 Graphical summary of the general procedures to study the absorption and translocation of ^{14}C-herbicides in weed and crop plants. The washing of the non-absorbed ^{14}C-herbicide and the assembly of the plants in filter paper is carried out at the different time intervals defined during the planning of the experiments, depending on the herbicide and the research objectives. Drying of samples up to constant mass may be carried out in an oven or at room temperature. Plants used for autoradiography can be reused for combustion and liquid scintillation spectrometry (LSS) analyses once ^{14}C-herbicide translocation has been visualized in a storage phosphor system.

close to the growth stage at which labeled rates of a trade herbicide are applied in the field (Nandula and Vencill 2015). Select the leaf to be treated with the ^{14}C-herbicide, preferably the youngest but fully expanded apical leaf, referred to as "treated leaf." Some researchers recommend doing a cold herbicide treatment on the plants prior to the hot herbicide treatment to mimic the field conditions and to not cause artificially high translocation due to high concentration in one plant part; therefore, the "treated leaf" is protected with plastic film, aluminum foil, or small paper envelopes (Nandula and Vencill 2015; Mendes et al. 2017b). However, in the literature there are quite a few examples where cold herbicide treatment is not done, and hot treatment is performed directly.

Preparation and application of the ^{14}C-herbicide solution. The ^{14}C-herbicide solution (^{14}C-herbicide + trade herbicide) is prepared as described in Section 5.1, and the radioactivity is checked before application as a safeguard against potential error in calculations or mixing (Nandula and Vencill 2015). This radioactivity corresponds to 100% theoretical Bq applied and will be used to estimate the mass balance. The radiolabeled solution is applied using a microsyringe or micropipette by applying the droplet(s) on the adaxial surface of the leaves that have been predetermined to receive the treatment (Figure 5.4). The volume and number of droplets (usually 1–10 droplets of 0.5–1.0 µL) to be applied depends on the herbicide (systemic or contact herbicide) and the radioactivity of the radiolabeled molecule (Mendes et al. 2017b).

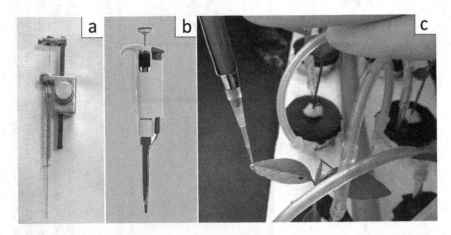

Figure 5.4 Microsyringe (a) and micropipette (b) (0.5–10 µL) used for application of ^{14}C-herbicide treatment solution. (c) Application of droplets of ^{14}C-imazamox solution with a micropipette on imazamox-resistant *Euphorbia heterophylla* leaves at the Laboratory of "Herbicidas en el Ambiente" of the Department of Agricultural Chemistry and Edaphology, UCO/Spain.

Plant handling after ^{14}C-herbicide application. At each harvest time, the treated leaf is rinsed with an adequate non-polar organic solvent (acetone, ethanol, or methanol) diluted in distilled water (Devine et al. 1984) to recover the non-absorbed ^{14}C-herbicide. The proportion of solvent must be established on preliminary tests with the studied molecule (Mendes et al. 2017b); for example, in De Prado's lab, the leaves treated with ^{14}C-glyphosate are washed three times with 1 mL of water-acetone (1:1 v/v). The rinsed solution from each wash is collected in 5-mL scintillation vials and mixed with scintillation liquid cocktail in the proportion 1:2 (v/v). After washing, the whole plants are carefully removed from the pot and the roots are washed with distilled water. Then, plants are sectioned into treated leaf, remainder of the plant (this plant section may be subsectioned into upper leaves, lower leaves, or stem according to the objectives of the research) and roots. Each plant tissue sample is stored in combustion cones and dried up to constant mass in an oven or at room temperature.

Combustion and LSS analyses. Because solids are not easily soluble in scintillation cocktails, dry plant tissues are combusted in a biological oxidizer (Figure 5.5). The [14]CO_2 released from the combustion is trapped in a mixture of a radioactive dioxide absorber and liquid scintillation cocktail suitable for ß energy. The liquid scintillation cocktails for rinsed solution and combustion and the radioactive dioxide absorber solution may be trade products or prepared in the laboratory. In De Prado's lab, an 18-mL mixture of Carbosorb and Permafluor (Perkin-Elmer) is generally used in the proportion 1:1 (v/v). Finally, the radioactivity recovered in the washes (non-absorbed ^{14}C-herbicide) and in the combustion of the samples (absorbed and translocated ^{14}C-herbicide in form of [14]CO_2) is quantified by LSS.

Statistical analyses and interpretation of data. Mass balance (percentage of recovery) and the absorption and translocation rates are calculated for each plant from the LSS data with the following equations:

$$\% \ recovery = \left(\frac{Bq \ from \ washes + Bq \ in \ treated \ leaf + Bq \ in \ plant + Bq \ in \ roots}{Bq \ total \ applied} \right) \times 100$$

$$\% \ absorption = \left(\frac{Bq \ in \ treated \ leaf + Bq \ in \ plant + Bq \ in \ roots}{Bq \ total \ applied} \right) \times 100$$

$$\% \ translocation = \left(\frac{Bq \ in \ plant + Bq \ in \ roots}{Bq \ in \ treated \ leaf + Bq \ in \ plant + Bq \ in \ roots} \right) \times 100$$

Absorption and translocation data can be analyzed by nonlinear regression analysis if the evaluation times are three or more times. If the absorption and/or translocation of two or more accessions are compared

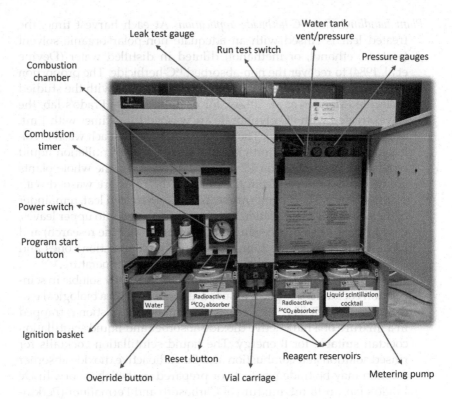

Figure 5.5 Automatic preparation and oxidation system (TriCarb 307, Packard Instrument Co.) from the Laboratory of "Herbicidas en el Ambiente" of the Department of Agricultural Chemistry and Edaphology, UCO/Spain, used for both single and dual radiolabeled samples containing ^{3}H and/or ^{14}C for use in liquid scintillation counter.

within an evaluation time, the data can be analyzed by analysis of variance, and when necessary, the mean differences can be tested using common mean separation procedures (Kniss et al. 2011).

Phosphorimaging. To visualize the ^{14}C-herbicide translocation, plants are treated and handled using the same media as described above. However, after removing the non-absorbed ^{14}C-herbicide by washing the treated leaf, the plants are preserved whole. Roots are carefully washed with distilled water and excess moisture removed with absorbent paper. After that, plants are fixed on filter paper (25 × 12.5 cm or the size of the PS film) and dried at room temperature or in an oven up to constant mass. Dry fixed plants are then placed side by side on a PS screen in the dark. The contact time (from 2 to 72 hours) of the plant with the PS screen depends on the plant species and the herbicide evaluated. Finally, translocation is revealed by scanning the ionizing radiation of the ^{14}C-herbicide absorbed in the PS screen in a PS system. Eventually, when resources are limited, the treated plants used for phosphorimaging are sectioned and combusted for LSS analyses. Figure 5.6 illustrates differences of ^{14}C-glyphosate translocation in *Cologania broussoneitti* plants, a tolerant species that retained the greatest amount of herbicide in the treated leaf and translocated only small amounts to young shoots, and a suscepti-ble *Conyza bonariensis* biotype, where ^{14}C-glyphosate translocation progressively increased from treated leaves to the remainder of the plant with the passage of time.

5.7.2 Herbicide metabolism

Radioactivity-based methods have been used to study herbicide metabo-lism by measuring total radioactivity; however, LSS and phosphorimaging do not provide information regarding herbicide metabolism because the radioactivity can come from a non-metabolic degradation that occurred outside the plant by photolysis (Rojano-Delgado et al. 2014). Therefore, even when radiolabeled herbicides are used, analytical methods are most proper for studying herbicide metabolism, consisting of three fundamental steps: plant preparation and herbicide application, extraction and separation, and identification of the herbicide and its metabolites (if any) (Figure 5.7).

The experimental design as well as plant preparation and herbicide application are similar to absorption and translocation experiments; there-fore, if ^{14}C-herbicide has been applied, the treated leaf must be washed with a non-polar solvent as described above. Fresh plant samples collected at defined times that cannot be processed immediately must be frozen in liquid nitrogen and stored at least at −40°C until use to ensure the stability of the active substances and metabolites (Mendes et al. 2017b).

Plant tissues are macerated with liquid N_2 in cold mortars, and the resulting fine powder is manually or mechanically homogenized with cold solvent at 80% (v/v) to carry out the extraction. It is important to know the proper extraction solvent system for each herbicide (Mendes et al. 2017b). The solution is centrifuged, the supernatant decanted, and the resi-due (pellet) is subjected to two additional extractions with the same cold solvent at 80% and 50% (v/v), respectively. Supernatants are mixed, and radioactivity is measured by LSS to determine the recovery percentage if ^{14}C-herbicide was used. Mixes of supernatant are evaporated with hot air, resuspended in 200 to 500 μL of solvent at 50%, and centrifuged. The final sample is analyzed with the most appropriate analytical method compat-ible with the herbicide molecule under study in comparison to analytical standards of the herbicide and its possible metabolites. Among the tech-niques commonly used in studies of herbicide metabolism are capillary

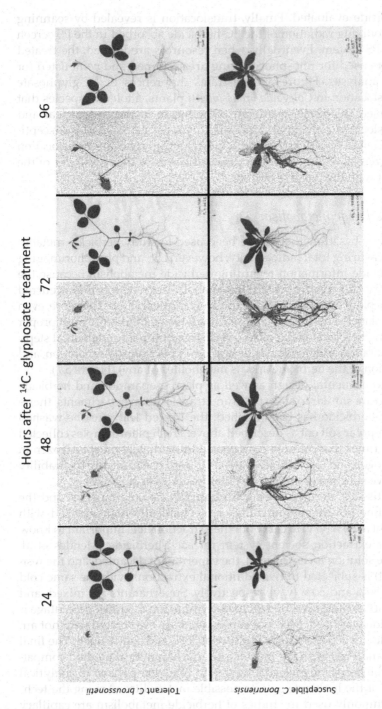

Figure 5.6 Autoradiographies and digital images of *Cologania broussonetti* and *Conyza bonariensis* plants, glyphosate-tolerant and -susceptible, respectively, from 24 to 96 hours after ¹⁴C-glyphsoate treatment. Plants were treated with 1-μL droplet of ¹⁴C-glyphosate solution at the concentration of 1 g ae L⁻¹ (200 g ae ha⁻¹) and a specific activity of 834 Bq plant⁻¹ (50,000 dpm). Filter paper assembly plants were beside a phosphor storage screen for 14 h before scanning on the Cyclone Plus phosphor storage imaging system (Alcántara-de la Cruz et al. 2016). Red zones of the autoradiographies correspond to the highest ionizing radiation from ¹⁴C-glyphosate.

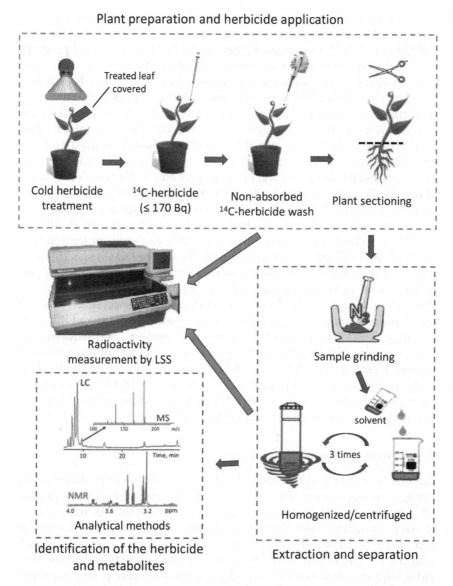

Figure 5.7 Graphical summary of the general procedures to study herbicide metabolism in weed and crop plants. Steps involving the use of ^{14}C-herbicides can be avoided in most cases; therefore, liquid scintillation spectrometry (LSS) analyses are not necessary.

electrophoresis (CE), gas chromatography (GC), high-performance liquid chromatography (HPLC), thin layer chromatography (TLC), depending on the herbicide molecule.

Detection and quantification of herbicide and metabolites can be improved by coupling to a mass spectrometry (MS) detector (Rojano-Delgado et al. 2014; Freud and Hegeman 2017). However, some potential herbicide target metabolites are not yet commercially available, which has delayed research on herbicide metabolism in resistant weeds. Another common drawback of these procedures is the need to establish exclusive extraction and identification methodologies, not only for the detection of the herbicide and its possible metabolites, but also in some cases for each plant species; i.e., it is possible that the methodology used to detect glyphosate in a monocot may not be effective in detecting this herbicide in a dicot. Establishing fast and effective procedures for the identification of metabolites without standards and without the use of radioactive compounds is a great challenge, since it is necessary to discover the behavior of a plant in relation to a certain herbicide (Rojano-Delgado et al. 2014).

An analytical approach that has been little explored for the study of herbicide metabolism is nuclear magnetic resonance (NMR) spectroscopy, a technique that makes it possible to take advantage of the labeling procedures used in MS-based metabolomic approaches with stable isotopes (Freund and Hegeman 2017). In vivo NMR offered a specialized probe of metabolism in living plant systems (d'Avignon and Ge, 2018), and this technique (by ^{31}P NMR) was essential to prove that rapid vacuolar sequestration may be responsible for glyphosate resistance in weeds that show reduced translocation (Ge et al. 2010, 2012, 2014). Therefore, developing methods to better understand not only herbicide-associated metabolic alterations in plants, but also to understand other NTSR mechanisms, will be important in planning new strategies for managing herbicide-resistant weeds.

5.8   Safe use of radioisotopes

Safety of the user and the general public is of utmost importance in the storage and application of radioisotopes; therefore, it is necessary to comply with regulations, laws, and guidelines regarding the safe use of radioactive materials (Palmer, 2019). Experiments with radioisotopes must be conducted in laboratories licensed by the regulatory agencies of each country. Personnel authorized to handle radioactive materials must have radiation protection training and are required to wear personal protective equipment (lab coat, gloves, gas mask, and goggles) exclusively for working with radioisotopes, made of easy-to-clean or disposable materials (Mendes et al. 2017b).

All rooms where radiolabeled material is handled or stored, as well as the equipment and materials used for its handling, must be duly identified with the international symbol of radioactive material, and only authorized personnel must be allowed access and use. If the laboratory works with different classes of radioisotopes (stable or unstable, or ionizing radiation α, β, χ, or γ) it is mandatory to assign different spaces or rooms for each type of isotope; i.e., the laboratory must have clearly identified radiological areas (monitored, controlled, limited stay, regulated stay, prohibited access) (Bowman and MacMurdo 1974). In addition, the laboratory must have a radiation detector (generally Geiger-Müller) that must be switched on when handling radiolabeled material, and the surface where it will be handled must be covered with an impermeable plastic film to avoid contamination of the equipment (Mendes et al. 2017b).

Radioactive waste is that material for which no utility is foreseen and that emits ionizing radiation due to being contaminated with radioisotopes; therefore, it needs to be stored and disposed of properly in order not to cause damage to health or the environment (Palmer 2019). Radioactive waste generation should be minimized by reducing volumes and streamlining operations. Research with ^{14}C-herbicides generates both solid and liquid waste in the form of liquid scintillation vials, disposable gloves, biological waste as well as the reagents in which $^{14}CO_2$ was trapped, the liquid scintillation cocktail, among others. These wastes must be adequately conditioned to allow their storage or transport and must be identified with all the information on the radioisotope, specific activity, volume, physical and chemical properties, generation date, and required storage period (Mendes et al. 2017b). It is important not to combine radioisotopes in the same container. The waste must be stored in a place designated for this purpose, suitably shielded, and located as far away as possible from the rest of the laboratory (Giusti 2009). Radioisotopes with a half-life of less than 100 d, such as ^{32}P, ^{33}P, ^{35}S, ^{125}I, can be discarded as conventional waste when its activity is below the exemption levels indicated in the corresponding legislation and the radiation levels correspond to the natural radioactive background. When the half-life exceeds 100 d, as in the case of ^{14}C and ^{3}H, radioactive waste can be temporarily stored, but then must be transferred to a radioactive waste management agency for final disposal (Bowman and MacMurdo 1974). For annual limits of exposure to ionizing radiation, as well as additional information on specific radioisotopes and procedures, check the annals of the International Commission on Radiological Protection (ICRP 2000), the *Handbook of Radioactivity Analysis* (L'Annunziata 2020), and the local reactive management guidelines of each country.

5.9 Concluding remarks

Herbicides must overcome a several barriers before reaching their target site; therefore, the main objective in designing herbicide formulations is to ensure that the active ingredient is retained, absorbed, and translocated in sufficient concentration to be lethal for weeds. At the same time, herbicide must be safe for crops, human health, wildlife, and the environment.

This chapter addressed the physical, physiological, biochemical, and in some cases even metabolic events that may prevent the herbicide from reaching its target site at lethal concentration, focused mainly on the NTSR mechanisms since they have gained great relevance in recent years by the considerable increase in cases of herbicide resistance around the world involving these mechanisms. NTSR mechanisms may affect various stages of the route of a herbicide, and the elucidation of these mechanisms is a great challenge, since they are governed by multiple genes that can be in any step of the herbicide route.

Absorption, translocation, and metabolism studies using ^{14}C-herbicides as tracers, as well as other analytical approaches, have been crucial in determining the location of the herbicide active ingredient in plant tissues and organelles, both for approval of new herbicide formulations and/or to elucidate NTSR mechanisms.

Due to the great usefulness of ^{14}C-herbicides, as they offer the possibility of determining very small amounts with precision, this chapter has described the general, but not arbitrary, theoretical and practical guidelines on the proper use of radiolabeled herbicides in the laboratory for the evaluation of the absorption, translocation, and metabolism of herbicides. Although these studies do not reveal which multiple genes are involved in herbicide resistance, they are still essential to affirm or rule out the participation of NTSR mechanisms in a given herbicide resistance case, and in some cases may help to dictate guidelines for future research.

Acknowledgments

RAC and MFGFS thank the "Fundação de Amparo à Pesquisa do Estado de São Paulo – FAPESP" for the financial support (Grants 2018/15910-6 and 2014/50918-7, respectively).

References

Alcántara-de la Cruz R, Barro F, Domínguez-Valenzuela JA, De Prado R. 2016. Physiological, morphological and biochemical studies of glyphosate tolerance in Mexican Cologania (*Cologania broussonetii* (Balb.) DC.). *Plant Physiology and Biochemistry*, 98: 72–80.

Barthe N, Maîtrejean S, Carvou N, Cardona A. 2020. High-resolution beta imaging. In *Handbook of Radioactivity Analysis: Radioanalytical Applications*, ed. M L'Annunziata (4th ed), 669–727. Elsevier: Academic Press.

Bowman WW, MacMurdo KW. 1974. Radioactive-decay gammas: Ordered by energy and nuclide. *Atomic Data and Nuclear Data Tables*, 13: 89–292.

Burkhardt J, Basi S, Pariyar S, Hunsche M. 2012. Stomatal penetration by aqueous solutions – an update involving leaf surface particles. *New Phytologist*, 193: 774–787.

Dayan FE. 2019. Current status and future prospects in herbicide discovery. *Plants* 8: e341.

d'Avignon DA, Ge X. 2018. In vivo NMR investigations of glyphosate influences on plant metabolism. *Journal of Magnetic Resonance,* 292: 59–72.

Délye C, Jasieniuk M, Le Corre V. 2013. Deciphering the evolution of herbicide resistance in weeds. *Trends in Genetics,* 29: 649–658.

Devine MD, Bestman HD, Hall C, Born V. 1984. Leaf wash techniques for estimation of foliar absorption of herbicides. *Weed Science,* 3: 418–425.

Devine MD, Duke SO, Fedtke C. 1993. Physiology of herbicide action. *Weed Technology.* 8: 418–419.

de Goeij JJM and Bonardi ML. 2005. How do we define the concepts specific activity, radioactive concentration, carrier, carrier-free and no-carrier-added? *Journal of Radioanalytical and Nuclear Chemistry,* 263: 13–18.

Dimaano NG, Yamaguchi T, Fukunishi K, Tominaga T, Iwakami S. 2020. Functional characterization of cytochrome P450 CYP81A subfamily to disclose the pattern of cross-resistance in *Echinochloa phyllopogon. Plant Molecular Biology,* 102: 403–416.

Domínguez-Mendez R, Alcántara-de la Cruz R, Rojano-Delgado AM, Silveira HM, Portugal J, Hipolito HEC, De Prado R. 2019. Stacked traits conferring multiple resistance to imazamox and glufosinate in soft wheat. *Pest Management Science,* 75: 648–657.

Duke SO, Dayan FE. 2018. Herbicides. In *Encyclopedia of Life Sciences.* Chichester: John Wiley & Sons, Ltd.

Duke SO. 2012. Why have no new herbicide modes of action appeared in recent years? *Pest Management Science,* 68: 505–512.

Feng PCC, Tran M, Chiu T, Sammons RD. 2004. Investigations into glyphosate-resistant horseweed (*Conyza canadensis*): retention, uptake, translocation, and metabolism. *Weed Science.* 52: 498–505.

Fernández-Moreno PT, Alcántara-de la Cruz R, Smeda RJ, De Prado R. 2017. Differential resistance mechanisms to glyphosate result in fitness cost for *Lolium perenne* and *L. multiflorum. Frontiers in Plant Science.* 8: 1796.

Freud DM, Hegeman AD. 2017. Recent advances in stable isotope-enabled mass spectrometry-based plant metabolomics. *Current Opinion in Biotechnology.* 43: 41–48.

Fujita M, Fujita Y, Iuchi S, Yamada K, Kobayashi Y, Urano K, Kobayashi M, Shinozaki KY, Shinozaki K. 2012. Natural variation in a polyamine transporter determines paraquat tolerance in *Arabidopsis. Proceedings of the National Academy of Sciences.* 16: 6343–6347.

Gaines TA, Duke SO, Morran S, Rigon CAG, Tranel PJ, Kupper A, Dayan FE. 2020. Mechanisms of evolved herbicide resistance. *The Journal of Biological Chemistry.* 295: 10307–10330.

Gaines TA, Patterson EL. Neve P. 2019. Molecular mechanisms of adaptive evolution revealed by global selection for glyphosate resistance. *New Phytologist.* 223: 1770–1775.

Gauvrit C. 2003. Glyphosate response to calcium, ethoxylated amine surfactant, and ammonium sulfate. *Weed Technology.* 17: 799–804.

Ge X, D'Avignon DA, Ackerman J. Collavo A, Sattin M, Ostrander EL, Hall EL, Sammons RD, Preston C. 2012. Vacuolar glyphosate-sequestration correlates with glyphosate resistance in ryegrass (*Lolium* spp.) from Australia, South America, and Europe: A ^{31}P NMR investigation. *Journal of Agricultural and Food Chemistry.* 60: 12432–12450.

Ge X, D'Avignon DA, Ackerman JJH, Sammons RD. 2014. In vivo ^{31}P-nuclear magnetic resonance studies of glyphosate uptake, vacuolar sequestration, and tonoplast pump activity in glyphosate-resistant horseweed. *Plant Physiology.* 166: 1255–1268.

Ge X, D'Avignon DA, Ackerman J. 2011. Rapid vacuolar sequestration: the horseweed glyphosate resistance mechanism. *Pest Management Science.* 66: 345–348.

Ghanizadeh H, Harrington KC. 2017. Non-target site mechanisms of resistance to herbicides. *Critical Reviews in Plant Sciences.* 36: 24–34.

González-Torralva F, Cruz-Hipolito H, Bastida F. Mulleder N, Smeda RJS, De Prado R. 2010. Differential susceptibility to glyphosate among the *Conyza* weed species in Spain. *Journal of Agricultural and Food Chemistry.* 58: 4361–4366.

González-Torralva F, Gil-Humanes J, Barro F, Brants I, De Prado R. 2012. Target site mutation and reduced translocation are present in a glyphosate-resistant *Lolium multiflorum* Lam. biotype from Spain. *Plant Physiology and Biochemistry.* 58: 16–22.

Grangeot M, Chauvel B, Gauvrit C. 2006. Spray retention, foliar uptake and translocation of glufosinate and glyphosate in *Ambrosia artemisiifolia.* *Weed Research.* 46: 152–162.

Gupta PK. 2011. Herbicides and fungicides. In *Reproductive and Developmental Toxicology,* ed. RC Gupta (1st ed), 503–521. Elsevier: Academic Press.

Giusti L. 2009. A review of waste management practices and their impact on human health. *Waste Management.* 29: 2227–2239.

Hart JJ, DiTomaso JM, Linscott DL, Kochian LV. 1992. Transport interactions between paraquat and polyamines in roots of intact maize seedlings. *Plant Physiology.* 99: 1400–1405.

Hess FD. 2018. Herbicide absorption and translocation and their relationship to plant tolerances and susceptibility. In *Weed Physiology. Vol. II: Herbicide Physiology.* ed. SO Duke (2nd ed), 191–214. Boca Raton: CRC Press.

Hess FD, Falk RH. 1990. Herbicide deposition on leaf surfaces. *Weed Science,* 38: 280–288.

Hull HM. 1970. Leaf structure as related to absorption of pesticides and other compounds. In *Residue Reviews.* ed FA Gunther and JD Gunther (1st ed), 1–150. New York: Springer.

International Commission on Radiological Protection (ICRP). 2012. Annals of the ICRP –Occupational Intakes of Radionuclides Part 1. Available from: https://www.icrp.org/docs/Occupational_Intakes_P1_for_consultation.pdf (accessed May 10, 2020).

Jiménez F, Fernández P, Rojano-Delgado AM, Alcántara R, De Prado R. 2015. Resistance to imazamox in Clearfield soft wheat (*Triticum aestivum* L.). *Crop Protection,* 78, 15–19.

Kniss AR, Vassios JD, Nissen SJ, Ritz C. 2011. Nonlinear regression analysis of herbicide absorption studies. *Weed Science,* 59: 601–610.

Kosma DK, Bourdenx B, Bernard A, Parsons EP, Lu S, Joubes J, Jenks MA. 2009. The impact of water deficiency on leaf cuticle lipids of Arabidopsis. *Plant Physiology,* 151: 1918–1929.

L'Annunziata M. 2020. *Handbook of Radioactivity Analysis: Radioanalytical Aapplications* (4th ed), 1047 p. Elsevier: Academic Press.

Mendes KF, Martins BAB, Reis FC, Dias ACR, Tornisielo VL. 2017a. Methodologies to study the behavior of herbicides on plants and the soil using radioisotopes. *Planta Daninha.* 35: e017154232.

Mendes KF, Silveira RF, Inoue MH, Tornisielo VL. 2017b. Procedures for detection of resistant weeds using ^{14}C-herbicide absorption, translocation, and metabolism. In *Herbicide Resistance in Weeds and Crops*, ed. Z Pacanoski, 159–176 (1st ed). London: IntechOpen.

Menendez J, Rojano-Delgado MA, De Prado R. 2014. Differences in herbicide uptake, translocation, and distribution as sources of herbicide resistance in weeds. In *Retention, Uptake, and Translocation of Agrochemicals in Plants*, ed. K Myung, NM Satchivi and CK Kingston (1st ed), 141–157. American Chemical Society.

Michitte P, De Prado R, Espinoza N, Ruiz-Santaella JP. 2007. Mechanisms of resistance to glyphosate in a ryegrass (*Lolium multiflorum*) biotype from Chile. *Weed Science*, 55: 435–440.

Moretti ML, Alárcon-Reverte R, Pearce S, Morran S, Hanson BD. 2017. Transcription of putative tonoplast transporters in response to glyphosate and paraquat stress in *Conyza bonariensis* and *Conyza canadensis* and selection of reference genes for qRT-PCR. *PLoS One*. 12: e0180794.

Moss S. 2017. Herbicide resistance weeds. In *Weed Research: Expanding Horizons*, ed. PE Hatcher and RJ Froud-Williams (1st ed), 181–214. Chichester: John Wiley & Sons, Ltd.

Nandula VK, Vencill WK. 2015. Herbicide absorption and translocation in plants using radioisotopes. *Weed Science*, 63: 140–151.

Nandula VK, Riechers DE, Ferhatoglu Y, Barrett M, Duke SO, Dayan FE, Cavalleri AG, Jones CT, Wortley DJ, Onkokesung N, Hicks MB, Edwards R, Gaines T, Iwakami S, Jugulam M, Ma R. 2019. Herbicide metabolism: Crop selectivity, bioactivation, weed resistance, and regulation. *Weed Science*, 67: 149–175.

Palma-Bautista C, Vazquez-Garcia JG, Travlos L, Tataridas A, Kanatas P, Valenzuela JAD, De Prado R. 2020. Effect of adjuvant on glyphosate effectiveness, retention, absorption and translocation in *Lolium rigidum* and *Conyza canadenses*. *Plants* 9: e297.

Palmer RA. 2019. Radioactive waste management. In *Waste: A Handbook for Management*, ed. TM Letcher and DA Vallero (2nd ed), 225–234. Elsevier: Academic Press.

Penner N, Klunk LJ, Prakash C. 2009. Human radiolabeled mass balance studies: Objectives, utilities and limitations. *Biopharmaceutics & Drug Disposition*, 30: 185–203.

Perkin-Elmer. 2020. Specifications of Cyclone Plus Storage Phosphor System. https://www.perkinelmer.com/CMSResources/Images/44-73873SPC_CyclonePlusStoragePhosphor.pdf (accessed May 10, 2020).

Reddy KN, Locke MA. 1996. Imazaquin spray retention, foliar washoff and runoff losses under simulated rainfall. *Pesticide Science*, 48: 179–187.

Rojano-Delgado AM, Menéndez J, De Prado R. 2014. Absorption and penetration of herbicides viewed in metabolism studies: Case of glufosinate and imazamox in wheat. In *Retention, Uptake, and Translocation of Agrochemicals in Plants*, ed. K Myung, NM Satchivi and CK Kingston (1st ed) 159–165. American Chemical Society.

Rojano-Delgado AM, Portugal JM, Palma-Bautista C, Alcántara de la Cruz R, Torra J, Alcántara E, De Prado R. 2019. Target site as the main mechanism of resistance to imazamox in a *Euphorbia heterophylla* biotype. *Scientific Reports*, 9: 15423.

Rosario JM, Cruz-Hipolito H, Smeda RJ, De Prado R. 2011. White mustard (*Sinapis alba*) resistance to ALS-inhibiting herbicides and alternative herbicides for control in Spain. *European Journal of Agronomy*, 35: 57–62.

Schreiber L. 2006. Review of sorption and diffusion of lipophilic molecules in cuticular waxes and the effects of accelerators on solute mobilities. *Journal of Experimental Botany*, 57: 2515–2523.

Shaner DL. 2009. Role of translocation as a mechanism of resistance to glyphosate. *Weed Science*, 57: 118–123.

Sharma MP, Van Den Born WH, McBeath DK. 1978. Spray retention, foliar penetration, translocation and selectivity of asulam in wild oats and flax. *Weed Research*, 18: 169–173.

Singh S, Kumar V, Datta S, Wani AB, Dhanjal DS, Romero R, Singh J. 2020. Glyphosate uptake, translocation, resistance emergence in crops, analytical monitoring, toxicity and degradation: A review. *Environmental Chemistry Letters*, 18: 663–702.

Sterling TM. 1994. Mechanisms of herbicide absorption across plant membranes and accumulation in plant cells. *Weed Science*, 42: 263–276.

Tani E, Chachalis D, Travlos IS. 2015. A glyphosate resistance mechanism in *Conyza canadensis* involves synchronization of EPSPS and ABC-transporter genes. *Plant Molecular Biology Reporter*, 33: 1721–1730.

Van Kirk C, Feinberg LA, Robertson DJ, Freeman WM, Vrana KE. 2010. Phosphorimager. In *Encyclopedia of Life Sciences*. Chichester: John Wiley & Sons, Ltd.

Vila-Aiub MM. 2019. Fitness of herbicide-resistant weeds: Current knowledge and implications for management. *Plants* 8: e469.

Weed Science Society of America (WSSA). 1998. Herbicide resistance and herbicide tolerance definitions. *Weed Technol.* 12: 789.

Yamaguchi S, Crafts A. 1958. Autoradiographic method for studying absorption and translocation of herbicides using C^{14}-labeled compounds. *Hilgardia* 28: 161–191.

Yao C, Myung K, Wang N, Johnson A. 2014. Spray retention of crop protection agrochemicals on the plant surface. In *Retention, Uptake, and Translocation of Agrochemicals in Plants*, ed. K Myung, NM Satchivi and CK Kingston (1st ed), 1–22. American Chemical Society.

Yu Q, Powles S. 2014. Metabolism-based herbicide resistance and cross-resistance in crop weeds: A threat to herbicide sustainability and global crop production. *Plant Physiology*, 166: 1106–1118.

Zawierucha JE, Penner D. 2000. Absorption, translocation, metabolism, and spray retention of quinclorac in *Digitaria sanguinalis* and *Eleusine indica*. *Weed Science*, 48:196–301.

chapter six

Radiological protection for the use of radiation and radioisotopes in agricultural research

Luca Ciciani[1], Alessandro Rizzo[2]
[1]*Istituto di Radioprotezione, Ente Nazionale per le Nuove Tecnologie, l'Energia e lo Sviluppo Sostenibile, Centro Ricerche ENEA Casaccia, Santa Maria di Galeria (RM), Italy*
[2]*Laboratorio di Sorveglianza Ambientale, Istituto di Radioprotezione, Ente Nazionale per le Nuove Tecnologie, l'Energia e lo Sviluppo Sostenibile, Centro Ricerche ENEA Casaccia, Santa Maria di Galeria (RM), Italy*

Contents

6.1   Introduction: use of radiation and radioisotopes in agricultural research

The use of radiation and radioisotopes in agricultural research is widespread and still increasing, as these methodologies have a wide range of applications. In particular, food and crop irradiation has been used for decades to achieve food sterilization, while radiation-induced genetic crop modification methods (Alam et al. 2001) are employed for producing new crop varieties with better resistance to water drought and bacterial infestation or other diseases (Osborne 2015). Insect sterilization by radiation is a tool used for pest control (Shalnov 1976, Lamm 1979). Radiation for these applications can be produced by high activity gamma sources or x-ray generators as well as particle accelerators or nuclear reactors. Another area of interest is the use of neutron radiation for studying the elemental composition of organic materials (de Olivera et al. 2013).

Radiotracers are powerful investigation tools in a variety of areas, including soil fertility, soil ion mobility, plant nutrient uptake and translocation (IAEA 2001), plant metabolisms, pesticide and herbicide behavior in soil and plants (crop and weed) (Nandula and Vencill 2015, Osborne 2015), and photosynthesis mechanisms (Alam et al. 2001). Radiotracers are also used in studies concerning the environmental fate of radioactive material as originating from nuclear accidents or radiological contaminations, nuclear power plant operations, radioactive discharges, short- and long-term disposal of radioactive waste (Brechignac et al. 2001, Weater et al. 2007), or from other origins.

One advantage of using radioactive tracers (or radiotracers) over chemical ones is that these have the same chemical behavior on their stable counterparts, but can be easily differentiated from ions already present in the compound or environment and so are able to provide information on the biochemical processes occurring (IAEA 2001, Mendes et al. 2017). Furthermore, radiotracers offer a greater sensitivity, a stepwise description of particular elements of the metabolic system, and are easily detected by autography on x-ray film or phosphorus blade images, and gamma counting/spectrometry or liquid scintillation (Mendes et al. 2017).

The Soil and Water Management and Crop Nutrition Section of the Joint FAO/IAEA Programme started in 1996 with the strategic objective to develop and promote the adoption of nuclear-based technologies for optimizing soil, water, and nutrient management in well-defined cropping systems and agro-ecological zones, which support intensification of crop production and preservation of natural resources. The program promoted the adoption and implementation of an integrated approach to soil, water, and nutrient management and the coordination of research in technical cooperation projects (IAEA 2001).

However, working with radiation and radioisotopes gave rise to extra hazards when working with conventional chemicals, which researchers and operators in agricultural research may not be accustomed to. Such hazards originate from the peculiar characteristics of the radiation phenomena and its interaction with unanimated and animated matters, the invisibility of radiation to human eyes, the very small mass amounts of radiotracers required to produce harm (in respect to normal chemicals), and the measuring methods involved.

When working with ionizing radiation, the workers' safety is ensured by a radiation protection system, a complex of authorizations, responsibilities, procedures, measurements, and instrumentation that satisfy a regulatory framework built to ensure the safety and health protection of workers, the population, and the environment. Such a system is created following radiological protection principles and studies, which are adopted by nations through a regulatory framework. A radiation practice, i.e. a facility using radiation sources, must conform to such regulatory framework and demonstrate the fulfillment of the stated requirements in order to perform its activity.

In this chapter, the aim is to provide researchers and operators in the agricultural research field who work with radiation sources an overview of the radiological protection principles, approaches, and methodologies used to ensure workers and the general population protection from radiation health risks. Given the wide range of applications, the chapter is necessarily quite general with respect to the effective development and implementation of a radiation protection system for a particular practice. Specific safety operational procedures for individual facilities are set up by the appointed radiation protection expert (RPE) on behalf of the authorization holder through a careful analysis of the risks and their relative mitigations and countermeasures.

In Section 6.2, an introduction to radiological protection is provided. Radiation types and sources and their main characteristics are discussed, and the radioactive decay law is presented. The principal physical quantities in use are then introduced, e.g. the concept of dose, and how these account for damage to living beings, which is divided into deterministic and stochastic effects. Radiological protection principles are introduced along with their practical application and implementation through regulations. The design and realization of a radiation protection system is then discussed, and more information is provided on the classification of areas and workers. Finally, in Section 6.2.6, the radiological protection of agricultural research facilities and laboratories is introduced, and separated in two macro areas – intense gamma or neutron irradiation facilities, and laboratories using radiotracers – since the risks and radiation protection systems are quite different for the two areas.

The practice of radiological protection for irradiation facilities is discussed in Section 6.3. Since for facilities in this area the key feature is radiation shielding, the principles for shielding design are introduced, and two case studies are presented. The first concerns a gamma irradiation facility for food sterilization, while the second a neutron irradiator allocating six americium-beryllium (Am-Be) sources.

The radiological protection for radiotracers practices is discussed in Section 6.4. The main characteristics of commonly used radiotracers are presented along with the radiotoxicity concept and grouping, as this one has implications on the type and characteristics of the laboratory needed for radionuclide manipulation. Information on laboratory design and requirements is then provided. General radiological protection guidelines and procedures are presented, with some details also for specific operations conducted in the laboratory. The environmental and worker monitoring programs are then discussed, including alarm and investigation level, and a method to evaluate the opportunity of establishing workers' internal contamination monitoring for specific radionuclides. In the final section, some information on the measurement equipment is provided.

The radioactive waste and its management are discussed in Section 6.5. The waste categories and the application of the exemption and clearance limit are presented along with information on the waste management strategy. The final section provides a collection of practical guidelines on management strategy and waste handling.

The final comments on the information presented and discussed, as well as on the radiation protection system and its implementation, are provided in Section 6.6.

6.2   *Radiological protection overview*

Radiological (or radiation) protection is a branch of the health physics science, and it is concerned with the impact caused by ionizing radiation to human health and the environment. Its ultimate aim is to protect workers and the general public from the dangers arising from the pacific use of radiation and radioactive substances. Radiological protection started to develop in the 1950s, responding to the demands of the expanding sciences and applications of the nuclear sector. If the nuclear sector represented the initial drive, decisive contributions were provided following the increasing application of radiation and radioisotopes in medicine for diagnostic, therapy, and research purposes.

Given the potentially high risks connected with the use of radiation and radioisotopes, this is controlled by national authorities through regulatory frames based on publications, directives, and guidelines issued by international reference bodies for radiological protection and use of nuclear technologies. The principal bodies that provide guidelines are listed below.

- *ICRP (International Commission for Radiological Protection)*: Issues reference publications concerning all aspects of radiation protection.
- *IAEA (International Atomic Energy Agency)*: Promotes, at a global level, the safe, secure, and peaceful use of nuclear technology.
- *ICU (International Commission on Radiation Unites and Measurements)*: Established international standards for radiation units and measurement.
- *NCRP (National Council on Radiation Protection and Measurement)*: A United States agency whose authoritative views and guidelines are recognized internationally.
- *EURATOM (European Atomic Energy Community)*: Responsible for studies, directives, and guidelines within the European community state members.

6.2.1 Ionizing radiation types, sources, and their characteristics

In physical terms, radiation is defined as the transport of energy with or without the displacement of matter. The ionizing radiation energy is usually measured in electron-volts (eV); that is, the energy of an electron in an electric field of 1 volt (1 eV = $1.6 \cdot 10^{-19}$ Joule), and its multiples (keV, MeV, GeV, etc.). Although that is a very small quantity in the macroscopic scale, it is a considerable amount of energy at the atomic scale.

There are two types of radiation: electromagnetic and corpuscular, i.e. made of subatomic particles with mass. Radiological protection focuses on ionizing radiation, i.e. radiation carrying enough energy to cause direct or indirect atomic ionization, which in turn causes, through a series of mechanisms, damage at cellular level and therefore to living organisms. Even if the smallest energy at which atomic ionization can occur is 3.89 eV (corresponding minimum energy needed to ionize the cesium atom), ionizing radiation is defined as that with frequency above $3 \cdot 10^{15}$ Hz (EURATOM 2013), corresponding to 12.4 eV. Therefore, for electromagnetic radiation, ionization starts with the far ultraviolet and higher energy, but the effect is more prominent at energy greater than 125 eV ($3.02 \ 10^{16}$ Hz), corresponding to the x-ray lower limit. Most international (EURATOM 2013) and national regulations apply only to radiation with energy above 5 keV ($1.2 \ 10^{18}$ Hz).

Any type of ionizing radiation, either electromagnetic or corpuscular, has its specific characteristics. The most common and relevant radiation types in the radiological protection context are electromagnetic high-energy radiation, i.e. x-ray/gamma radiation, and beta, alpha, and neutron radiations. Radiation is also associated with energetic protons, heavy ions, and other subatomic particles (muons, pions, etc.), but their use is far less common and to date not applied, to the authors' knowledge, in the agricultural research field, and therefore these are not further discussed.

Radiation is actually ubiquitous in the environment, originated by the sun and the stars (cosmic origin) or by primordial (uranium and thorium decay chains, and ^{42}K) and cosmogenic (mainly ^{7}Be, ^{14}C, and ^{24}Na) radionuclides. The average annual dose to humans due to the natural radiation background amounts to about 2 mSv. Other sources of radiation, as particle accelerators, nuclear power plants, and anthropogenic or artificial radionuclides (e.g. transuranic element, ^{137}Cs, ^{60}Co, ^{32}P, and many others) accounts for only a small fraction of the average annual dose.

The different radiation types and their characteristics are listed below.

X and gamma radiation: High-energy electromagnetic radiation, constituted by photons with energy from 125 eV upward. This corresponds to a frequency of $3.02 \ 10^{16}$ Hz, while visible light has frequency in the range of 430 to 770 10^{12} Hz. It is produced in alpha and beta decay of atomic nuclei (e.g. ^{60}Co and ^{241}Am sources), radiation generator as x-ray tubes, medical and research accelerators (gamma radiation), and nuclear reactors. Physically, x-rays are produced in atomic transitions, while gamma rays have a nuclear origin. Conventionally, electromagnetic radiation with energy of 100 keV and above is referred to as gamma radiation. Highly penetrating but low biological damage as energy is released and ionization produced over relatively "large" volumes. It is one of the main contributors to external exposure but also important for internal exposure.

Beta radiation: Made of electrons; it is produced by the beta decay of light nuclei (and often emitted together with gamma radiation) and by electron accelerators (e.g. medical and research). Moderate penetrating power, moderate damage, causes ionization in moderately small volumes. It is important for external exposure of skin, eye lens, and extremities, and internal exposure.

Alpha radiation: Made of helium nuclei (two protons and two neutrons); it is produced by the alpha decay of heavy nuclei and often emitted together with gamma radiation. Low penetration power, but highly damaging, as a large amount of energy is released in small volumes. For external exposure, it may be important cases of skin contamination, but extremely important for internal exposure.

Neutron radiation: Made of neutrons. Penetration power and damage depend strongly on the neutron energy. Neutrons are divided according to their energy, and the principal energy ranges are, in order of increasing energy: cold, thermal, epithermal, slow, intermediate, fast, and ultrafast. Produced by neutron radio-isotopic sources (e.g. Am-Be), particle accelerators (above 10 MeV), and nuclear reactors. Important for external exposure, but effective impact is strongly dependent on neutron energy. Internal exposure is not relevant, since (besides ^{242}Cf) radionuclides do not spontaneously emit neutrons in their decay.

Radiation sources can be divided in radiation generators, e.g. particle accelerators and x-ray tubes, nuclear reactors, and radionuclides. The main difference is that any generator emits radiation only when turned on (a low amount of radiation can be still produced by activated materials), while radionuclide sources emit radiation until they transform, through one or more radioactive decays, to stable elements. Nuclear reactors have characteristics of both: these produce a very large amount of radiation when the controlled chain reaction takes place (reactor on), and a lower, but still large, radiation amount is produced by the decay of radionuclide (fission and activation products) contained in the reactor when the chain reaction is stopped (reactor off).

Particle accelerators and nuclear reactors produce a complex radiation field, characterized by several particle types and energy. Particle flux and energy fluence (particle or energy crossing the surface unit in the time unit) are used to express the radiation field intensity.

On the other hand, radionuclide sources are still characterized by the emitted radiation type and energy, but their intensity is expressed by the activity, i.e. the amount of decay per second. From the radiological protection point of view, their physicochemical form plays an important role.

A radioisotope, or radionuclide, is just an unstable nuclear state (excess/deficit of neutrons) of a normally stable natural element. It spontaneously decays, emitting radiation, to other radioelements and ultimately to a stable nuclide. The entire decay sequence is called decay chain. The radioactive decay follows a simple exponential law, that is:

$$A(t) = A_0 e^{-\lambda(t-t_0)}$$

where $A(t)$ is the activity a time t, A_0 is the activity a time t_0, and λ is the decay constant, given by

$$\lambda = \frac{ln2}{T_{1/2}} = \frac{0.693}{T_{1/2}}$$

where $T_{1/2}$ is the radionuclide half-life, i.e. the time in which half of the initial nuclei decays.

For radionuclides in decay chains, the equations expressing their activity are more complicated and are provided by the Bateman equations (Bateman 1932), which account for both the decay and growth.

The activity is the number of decays that occurs in the time unit, measured in *Becquerel* (1 Bq = 1 decay per second). The conversion from Bq to *disintegration per minute (dpm)*, a unit also commonly used e.g. in liquid scintillation, is 1 Bq = 60 dpm, or 1 dpm = 0.01667 Bq. An older unit, rarely but still used sometimes, is the *Curie (Ci)*. The conversion from Curie to Becquerel is:

$$1\ Ci = 3.7\ 10^{10}\ Bq = 37\ GBq$$

Another important quantity for radioisotopes is the *specific activity* A_s, the activity for mass unit of a pure substance. This is calculated as

$$A_s\left(\frac{Bq}{g}\right) = \frac{\lambda N_{av}}{A_w}$$

where N_{av} is the *Avogadro's number*, equal to $6.02\ 10^{23}$ atoms/mole, and A_w is the *atomic weight* in g/mole. These quantities are tabulated for known radioemitters (Laboratoire National Henri Becquerel (LNHB 2020) and National Nuclear Data Center (NNDC 2020)).

In each transformation (i.e. decay) the radioisotopes emit one or more radiation types, with one or more energies, each one with its own specific probability (i.e. *yield*). The yield values range from 100% to very low values (e.g. 0.001%). The radiation emitted in each decay can potentially cause damage and it is therefore considered for radiological protection purposes.

An important classification for radionuclide sources is in sealed or unsealed:

- "Sealed source means a radioactive source in which the radioactive material is permanently sealed in a capsule or incorporated in a solid form with the objective of preventing, under normal conditions of use, any dispersion of radioactive substances" (EURATOM 2013).

- Any other source is defined as unsealed. Being without capsule, it bares the risk of being disperse in the environment. They can have a different physical state: solid, liquid, and gaseous (EURATOM 2013).

6.2.2   Radiation damage and dose

Radiation damage is proportional to the amount of radiation absorbed in the organs, living tissue, and the whole body. Without going into the mechanisms causing damage at a cellular level and then to organs and tissues, in radiological protection the radiation effects are divided in two main types: deterministic and stochastic (ICRP 1990, ICRP 2007). For deterministic effects, their entity is directly proportional to the amount of radiation absorbed, and it manifests on the exposed individual (with little delay), while stochastic effects occur with a probability proportional to the amount of radiation absorbed, and can manifest not only on the subject exposed (with a delay time that can reach decades), but also on their progeny.

Living tissues can be exposed to radiation from external sources, as so-called external irradiation, but also from internal sources when radionuclides are incorporated in the body. Incorporation can happen through inhalation, ingestion, skin absorption, or open wounds. The first two types of incorporation are the most considered in normal occupational situations, not just because these provide the highest incorporation, but also because the other two types are normally prevented by wearing personal protective equipment (PPE), e.g. laboratory coats, gloves, and avoiding cuts or wounds. However, the latter two can become important in specific situations, e.g. worker's accident or incident, when these need to be properly considered.

The impact of radiation on living organisms is quantified by several dosimetric protection quantities, which all have "dose" in their names. The correct one to be used depends on which type of exposure is considered and if the effects are deterministic or stochastic (ICRP 1990, ICRU 1998).

For deterministic effect, the quantity correlating radiation and damage is the *absorbed dose (D)*, measured in *Grey* (1 Gy = 1 Joule/kg). This accounts for the energy released onto living tissue/organs per mass unit.

For stochastic effect, two quantities are defined. The *equivalent dose* (H_T) is used for exposure of specific organs and tissues, such as skin, extremities, and eye crystalline lens. It accounts for the energy released in the tissue and the type of radiation (through *radiation weighting factors*), as this determines how such energy is released. The *effective dose (E)* accounts for the whole-body exposure, and in addition to the factors above, it also accounts for the different tissue radio-susceptibility (through *tissue weighting factors*). For internal exposure, the *committed effective dose (E(τ))* is used.

This accounts for the factors discussed above, and it is integrated over a period of 50 years for adults (or 70 years for infants or children). Given the radionuclide and the amount incorporated, the damage also depends on how this is distributed within the body, e.g. where the damage is actually done. This is accounted for by the *committed dose coefficient e(50)* (ICRP 1997), computed for both ingestion and inhalation. The doses for stochastic effects are all measured in *Sievert* (1 Sv = 1 Joule/kg).

In general, deterministic effects arise when individuals are exposed to large amounts of radiation, as in accidents or malfunctioning. This type of effect appears above a certain exposition threshold. Normally, in occupational activities (i.e. working with radiation sources) only the stochastic effects are considered, which have a zero-threshold limit, meaning that, in a precautionary way, it is assumed that even the smallest dose has, statistically, a non-zero probability to cause some damage. In reality, since the two types of effects are predominant at different exposition levels, only one is normally considered in function of the exposition level. However, in specific, e.g. intermediate exposure situations, it may be necessary to consider and account for both.

Returning to radiation, gamma radiation causes both external and internal irradiation. In particular, it is important to evaluate its impact on the whole body. Beta radiation is mostly concerned with irradiation to the extremities, skin, and crystal eye lens, and for internal irradiation. Alpha radiation external irradiation is usually negligible, as it travels very short distances (about 7 cm in air), but it can become important in the case of skin contamination. This type of radiation is definitely very important for internal irradiation, as it releases its energy in a very localized manner, e.g. within the organ/tissue where it is absorbed. Since, as discussed above, the incorporation takes place mostly through inhalation, this is the exposure pathway that deserves the greatest consideration. Neutron radiation is a strong contributor to external irradiation but, since radionuclides, besides ^{242}Cf which undergoes spontaneous fission, do no emit neutrons in their decay, internal exposure normally is not relevant for neutron radiation.

6.2.3 Radiological protection: basic principles, hazards, and risk management approach

In radiological protection, the radiation risk for workers, the general population, and the environment is managed according to three general principles (EURATOM 2013, ICRP 2007), which can be stated as:

> **Justification**: "The individual or societal benefit resulting from the practice outweighs the health detriment that it may cause" (EURATOM 2013).

Optimization: The dose amount should be reduced at the lowest possible level, according to the *ALARA* principle (*As Low As Reasonably Achievable*), where "reasonably achievable" takes into account technical, economic and societal factors (EURATOM 2013).

Dose limitation: "In planned exposure situations, the sum of doses to an individual shall not exceed the dose limits for occupational exposure or public exposure" (EURATOM 2013). The annual effective dose limits, as set for the population and classified workers, are summarized in Table 6.1 (EURATOM 2013). For the non-classified workers, the limit values are the same as for the population.

Effective dose limits for exposed workers can be increased in special situations (specified by national regulations) up to 50 mSv in a single year upon authorization from the competent authority. However, the total effective dose over 5 consecutive years, including the year in which the limit was increased, must not exceed 100 mSv.

In radiological protection, three principles work together: the risks are carefully assessed, evaluated, and managed to ensure health protection for workers and population.

The application of the justification principle is realized by a regulatory frame that requires that practices, depending on the risk involved, progressively require greater control by the authorities. In a graded approach, there are three regulatory ranges, depending on the total radionuclide activities and concentrations held or the radiation energy used in the planned installation. The boundaries for the three ranges are set by national authorities. Below the so-called exemption levels, no authorization is required (*exemption regime*). For practices in the second range (*notification regime*), it is necessary to inform the authority in charge of the planned activities. Above the notification values, the practice must submit

Table 6.1 Dose limit for population and exposed workers

	Whole body (mSv)	Skin (mSv)	Extremities (mSv)	Eye lens (mSv)
Population	1	50	–	15
Exposed worker	20	500	500	150 (20)[a]

[a] The limit of 150 mSv/y for the eye lens, established in the European Council Directive 29/96/EURATOM (EURATOM 1996) is still used by most countries. However, such limit was lowered to 20 mSv/y in the most recent European Council Directive 59/2013/EURATOM (EURATOM 2013).

a request for authorization and wait until this is granted before starting its operation (*authorization, or licensing, regime*).

In Europe, national regulations normally refer to the ranges specified by EURATOM in the European Council Directive 96/29/EURATOM (EURATOM 1996) and more recently in 2013/59/EURATOM (EURATOM 2013). A practice is in exemption regime if it has not radiological relevance, which means that due to its operation the individual of the most exposed group will not receive annually an effective dose larger than 10 μSv/y and the *collective effective dose* is no more than 1 Sv·man/y. This last quantity is the sum of the doses over the population that has been exposed. The criteria are expressed through operational limits for the total activity and activity concentration of the radionuclides held at the practice, and the radiological non relevance is proved if either the total activity or the activity concentration are lower than the respective limits. In 2013/59/EURATOM (EURATOM 2013), the concentration limits are listed in Annex VII, Tab. A, Part 1 (for artificial radionuclides) and the activity limits in Tab. B (column 3). The concentration values in column 2, Tab. B can be used in cases of a moderate amount of total activity, specified by the member state for the different types of practice. The exemption concentration values used by EURATOM are those provided by the IAEA (IAEA 2004).

If different radionuclides are held, or planned to be held, at the practice, the exemption is evaluated by

$$\sum_i \frac{A_i(Bq)}{A_i^{ex}(Bq)} \leq 1 \ OR \ \sum_i \frac{C_i\left(\frac{Bq}{g}\right)}{C_i^{ex}\left(\frac{Bq}{g}\right)} \leq 1$$

where i refers to the ith *radionuclide*, $A^{ex}(Bq)$ and $C^{ex}(Bq/g)$ are respectively *the activity* and *activity concentration exemption values*, and $A_i(Bq)$ and $C_i(Bq/g)$ the *activity* and *activity concentration* of the ith radionuclide held at the practice.

In this context, it is important to note that practices in the exemption regime must still respect the justification principle, so the benefit must offset the harm that radiation may cause. Therefore, practices not justified include those "involving addition of radioactive substances to food and beverages, frivolous use of radiation or radioactive substances in commodities or products" (e.g. toys, adornments, jewels) (IAEA 1996).

The application of the optimization principle is realized by using a system of devices, procedures, authorizations, and working behaviors that ensure that the dose absorbed by the workers during the operations follows the ALARA principle. Even if the system can be complex, the physical principles are simple: maintaining a certain distance between operator and source, reducing the time the operator is in the presence of the source, and using shielding barriers to attenuate the radiation reaching the worker and equipment to reduce possible radionuclide intake. In relation to internal exposure, this means that the sources must be kept physically confined, operations with unsealed sources must be performed under controlled atmosphere (e.g. fume cupboard, laboratory-controlled pressure), airborne activity is reduced by ventilation systems, and PPE (e.g. laboratory coats, gloves, gas mask, etc.) is used to further reduce the risk.

The dose limitation (third) principle is fulfilled by carrying out a dose evaluation prior to potential exposure, which makes it possible to adopt the correct strategy to minimize exposure and plan the control measurements to be used. Compliance is achieved by monitoring both the working environment and the individuals. Controls are performed by portable and fixed instrumentation, as radiation monitors (e.g. radiometers), by personal dosimeters, and by carrying out contamination surveys for working areas, atmospheric particulate and PPE, and periodic control of workers' radionuclide incorporation.

National regulations are built using these three principles, and the practices in the notification and authorization regime must fulfill the regulation requirements by designing and implementing an appropriate radiation protection system, discussed in the next section.

6.2.4   The radiation protection system

The radiation protection system is a complex of authorizations, procedures, guidelines, measurements, equipment, training, knowledge, and records built to ensure that risks related to the use of ionizing radiation are properly controlled, monitored, minimized, and acted upon. This is defined, implemented, and managed according to the national regulation framework. For the specific practice, several aspects are considered: the radiation type, its sources and intensities, the operations performed, the structure where those operations are performed, the workers and workplace measurements, the measuring equipment, the safety procedures implemented, and so forth.

The radiation protection system covers the entire lifespan of the practice: from the start, e.g. in the notification/authorization process, throughout the practice operational life, and even after the practice closes, as also during the closure and post-closure phase certain requirements must be fulfilled, e.g. documentation must be kept available to authorities for several years.

Among the different figures that contribute to this, including the license holder and the workers themselves as well as external services (e.g. medical and dosimetry service), a key role is held by the RPE, an individual whose competence in the field is recognized by the national authority.

The RPE acts on behalf of the license holder in managing several aspects of such system, and ultimately provides the necessary information and guidance to fulfill the regulatory requirements. In practical terms, the RPE acts as an advisor to the license holder for the aspects concerning with radiological protection listed in Art. 82 of the European Council Directive 2013/59/EURATOM (EURATOM 2013), and reported below:

- *Optimization and establishment of appropriate dose constraints*
- *Plans for new installations and acceptance into service of new or modified radiation sources (in relation to any engineering controls, design features, safety features, and warning devices relevant to radiation protection)*
- *Categorization of controlled and supervised areas*
- *Classification of workers*
- *Workplace and individual monitoring programs and related personal dosimetry*
- *Appropriate radiation monitoring instrumentation*
- *Quality assurance*
- *Environmental monitoring program*
- *Arrangements for radioactive waste management*
- *Arrangements for prevention of accidents and incidents*
- *Preparedness and response in emergency exposure situations*
- *Training and retraining programs for exposed workers*
- *Investigation and analysis of accidents and incidents and appropriate remedial actions*
- *Employment conditions for pregnant and breastfeeding workers*
- *Preparation of appropriate documentation such as prior risk assessments and written procedures*

The RPE will also liaise, where appropriate, with the occupational, medical, and dosimetry services for the area of competence.

6.2.5   Classification of areas and workers

One of the most important features for managing radiation risks is that workers and operational areas are classified according to potential exposure.

Exposed workers are classified in *two categories, A and B*. For category A (Cat. A) workers, the dose limits are those reported in Table 6.1, while for category B (Cat. B) the annual dose limits are about one-third of those; namely 6 mSv/y for effective dose, 15 mSv/y for equivalent dose for the eye lens, and 150 mSv/y for skin and extremities (EURATOM 2013). Different exposure limits bear different requirements for dose assessment. For Cat. A classified workers, this must be based on data from personal dosimeters and internal contamination exams, while for Cat. B it is possible to use only environmental measurements, even if personal data can also be used. The different risk entity also reflects on the depth and frequency of the medical surveillance. For instance, in Italy the medical surveillance of Cat. B workers is once a year, and twice a year for Cat. A.

The workplace area where radiation sources are used shall be classified on the basis of expected annual dose, and probability and magnitude of potential exposure (EURATOM 2013). The classification is in *supervised areas*, where the expected annual dose exceeds the limit for members of the public but not those for Cat. B workers, and *controlled areas*, where these latter limits are expected to be exceeded (but not the general limits of Table 6.1). Classified areas must be clearly indicated through proper signs reporting the source and risk type, have restricted access, and must undergo radiological surveillance, and working instructions pertinent to the radiological risk, sources, and the operations involved, shall be stated (EURATOM 2013). From the practical point of view, the main differences between the two types of classified areas are in the frequency, type, and extent of safety measurements undertaken.

Monitoring programs to evaluate the effective dose in the working area and for workers are discussed in detail in Sections 6.4.4 and 6.4.5.

6.2.6   Radiological protection for radiation and radioisotope use in agricultural research

Concerning the use of radiation and radioisotopes in agricultural research, two main macro areas can be distinguished: the use of high-intensity radiation for food sterilization/preservation, radiation-induced crop mutation, pest control/male insect sterilization (Shalnov 1976, Lamm 1979), and neutron radiation for elemental composition of organic material studies (de Olivera et al. 2013), and the use of radioisotopes as tracers in fertilizer/plant nutrition uptake studies, in pesticides, and herbicide behavior in soil and plant research, and studies on plant metabolism (Lamm 1979, Alam et al. 2001, IAEA 2001, D'Sousa 2014, Mendes et al. 2017). The risks, and therefore the radiation protection systems, for the two areas are quite different.

Practices belonging to the first area use intense gamma irradiation (intensity depends on the specific use), produced by sealed, high-intensity radioisotope sources as ^{60}Co and ^{137}Cs, or by particle accelerators or x-ray generators, and neutron radiation produced by radioisotopic neutron source (e.g. the typical Am-Be source) or nuclear reactors. The radiological protection issues in this field are mostly concerned with appropriate shielding, workers' external exposure, source safety and security, and air activation for neutrons with energy above 10 MeV. The radioactive waste produced essentially is made of disposable PPE, such as gloves, laboratory coats, and so forth, all with none or minimal activity. An important aspect regarding radiation application in this area, even if not concerned with workers safety, is the control of irradiation given to the crops/insect/ seeds, and the uniformity on the irradiation field.

On the other hand, practices in the second area deal essentially with radioisotope open sources, of low activity, which are manipulated and transferred to soil and plants. Radiochemical specific compounds are mostly commercially available and bought from dedicated agro-technical or agro-chemical companies; therefore, radiological protection issues associated with their production are not examined in this context. Radiological protection is essentially concerned with open/ unsealed source control, radiation hazards for radioisotope manipulation, correct laboratory design and settings, operations performed in fume cupboards, workers' internal contamination, and surface and laboratory equipment contamination and decontamination. Besides personal protective devices, radioactive waste is produced by conducting experiments with plant and soil, and even if in small quantity, it must be disposed of in the appropriate manner and according to national regulations.

The radiological protection aspects for practices in the two areas are discussed separately in Sections 6.3 and 6.4, while issues with radioactive waste management are discussed in Section 6.5.

6.3   Use of high dose radiation and shielding design

In agricultural research, the techniques involving strong irradiations, based on photonic and neutron radiations, are used in food/ crop sterilization and elemental composition research. The main concern, from the radiological protection point of view, is the design of the facility radiation shielding necessary to keep workers and general public exposure below the required limits, and also to minimize exposure according to the ALARA principle. Risk assessment in shielding design takes into account both types of radiation effects: it eliminates the risk related to deterministic effects and lowers that related to the stochastic effects.

The application of the ALARA principle to shielding design does not mean delivering a zero dose, but the lowest dose achievable considering economics and social factors. Specifically, using an "infinite" amount of shielding material to deliver a zero dose is not necessary, since individuals are exposed to natural radioactivity. It is needlessly expensive as cost associated with shielding would increase without leading to real benefits, and space demanding, which can pose a problem because large space is often not available.

Photons (energy in the order of MeV) and electrons (energy less than 10 MeV) are generally used for food/crop sterilization. The facility shielding design, i.e. barrier thickness and material used, depends on the radiation type and energy, because the different interaction processes occurring between radiation and matter imply that particles release their energy to the materials with different mechanisms.

Besides shielding materials, the key factors for maintaining exposure below the limits are limiting the exposure time and increasing the distance from the source. The effective source position in the irradiation room, and therefore its effective distance from the walls, and the occupancy factor in adjacent rooms both contribute to determine the shielding factor to be achieved.

For particle accelerators emitting gamma radiation, the amount of radiation that crosses the shielding, or barriers, is expressed by the *transmission factor* B_x, given by (NCRP 2005a)

$$B_x = \frac{P \cdot (d_{pri})^2}{W \cdot U \cdot T}$$

where P *(mSv/week)* is the weekly dose required after the barrier (normally at 30 cm from the external wall), $d_{pri}(m)$ is the distance between the source (e.g. accelerator head), W is the *machine workload factor* which accounts for how much radiation is produced weekly by the machine usage, U is the *barrier usage factor* (from 0 to 1), and T (between 0 and 1) is the *occupational factor* for the area after the barrier.

The workload factor also depends on the particle fluence and energy: accelerators, x-ray generators, or high activity sources used (e.g. in food/crop sterilization) have load factors considerably higher than those related to scientific research (e.g. x-ray diffractometry), due the large dose delivered (in the order of kGy or higher) for food sterilization. Higher workloads require lower transmission and therefore thicker barriers to reduce the delivered dose.

For gamma radioisotopic sealed source, the transmission factor B_{seal} is expressed as

$$B_{seal} = \frac{P \cdot (d_{pri})^2}{\dot{K} \cdot W \cdot U \cdot T}$$

where $\dot{K}$ (Gy m^2/week) is weekly *kerma rate in air* at 1 m from the source.

For gamma radioisotopic unsealed source, the transmission factor B_{un} is expressed as

$$B_{un} = \frac{P \cdot (d_{pri})^2}{\Gamma \cdot A}$$

where A(Bq) is the source activity and Γ is the *specific gamma constant*, which for each gamma emitter provides the dose rate (mSv/h) for a source of 1 Bq activity at 1 m from it.

The barrier thickness is normally expressed in number of *HVL* (*half value layer*) or *TVL* (*tenth value layer*), respectively – the material thickness necessary to attenuate the radiation by a factor two or ten. If N_{10} is the number of TVL, this is obtained as:

$$N_{10} = \log_{10} B_x$$

and the actual thickness S as:

$$S = TVL_0 + (N_{10} - 1) \cdot TVL_e$$

where TVL_0 and TVL_e are the *first* and *equilibrium TVL*, slightly different because the first needs to account also for the shielding of the built-up radiation.

The effective TVL values for different materials depend on the radiation type and energy, and for particle accelerator it can be found in several publications, such as *Radiation Protection for Particle Accelerator Facilities*, NCRP Report No. 144 (NCRP 2005a) or *Structural Shielding Design and Evaluation for Megavoltage X- and Gamma-ray Radiotherapy Facilities*, NCRP Report No. 151 (NCRP 2007).

For beta radiation it is customary to define the *maximum range* R_{max} (g/cm^2) as the deepest penetration of electron in a material. The range R_{max} can be calculated by several semi-empirical relations, and one of the most used is due to *Katz and Penfold* (Katz and Penfold 1952):

$$R_{max} = 0.412 \cdot E^{1.265 - 0.0954 \ln E} \text{ for } 0.01 < E_\beta \leq 2.5 \text{ MeV}$$

$$R_{max} = 0.530 \cdot E - 0.106 \text{ for } E_\beta > 2.5 \text{ MeV}$$

where the E_β (MeV) is the maximum beta energy in MeV. Dividing R_{max} by the *material density* ρ (g/cm^3), the effective thickness S (cm) is obtained.

The formulas above offer an indication of the main features for the shielding strategy. However, for accurate shielding design, it is necessary to refer for further details to international publications such as those from the NCRP (NCRP 2005a, NCRP 2005b, NCRP 2007). Even if these studies refer to the use of radiation in the medical field, they have a broad validity and therefore are applicable to the use of radiation in agriculture research. The formulas for neutron are not presented here, as these would require a lengthy discussion, but they can be found in the NCRP publications (NCRP 2007). However, it must be noted that neutron radiation is often associated with intense gamma radiation fields, and these both need to be accounted for.

The shielding materials used for photons and neutrons are different: photons (gammas or x-rays), after processes such as Compton scattering (e.g. diffusion) and pair (e^-–e^+) creation, are absorbed through photoelectric effect, which has a greater probability of occurring in high-Z (Z is the *atomic number*) materials, e.g. lead; neutrons must first lose most of their kinetic energy in elastic scattering processes (energy lost is more efficient in scattering with light nuclei such as hydrogen), in a process called *thermalization*, and then be absorbed by elements like cadmium, boron, or gadolinium, which have large cross-sections (i.e. probability) for neutron capture processes.

It is worth noting that an important multipurpose shielding material, used for both photons and neutrons, is concrete. This is due to its structural properties (e.g. used for bunker construction), low cost, and effectiveness in photon and neutron attenuation. Heavy aggregates, such as barite or iron ore, are used to produce concretes of higher density for special shielding purposes. Even if more expensive and difficult to pour for ensuring uniform composition and density, they possess a higher attenuation capability than ordinary concrete for gamma and neutron radiations, and can be used where needed (e.g. low space availability or need for high attenuation).

The remainder of this section focuses on two applications: the use of gamma radiation for food/crop sterilization (high delivered doses) and the use of thermal neutron for elemental analysis of samples. These make it possible to go into greater detail with the shielding design, and for the second case the results of numerical simulation for shielding neutron ^{241}Am-^{9}Be sources, used in an irradiator for research, highlight the physical key feature for neutron radiation shielding.

6.3.1   Considerations on shielding design for food/crop sterilization facilities using gamma radiation

Facilities for food/crop sterilization generally use high activity gamma-ray sources, in particular ^{60}Co (E_γ=1,17 MeV e E_γ=1,33 MeV) and ^{137}Cs (E_γ=660 keV), or electron accelerators to produce gamma beams (E_γ <5 MeV) through the bremsstrahlung (braking) effect, and the use of electron beams (E_e<10 MeV) is also documented (Boniglia et al. 2003). The delivered dose can be *low* (<kGy), *medium* (between 1 and 10 kGy) or *high* (>10 kGy), depending on the food/crop treatment required. The irradiation room (sometimes called *hot room*) is normally a concrete bunker where the food/crop is transported on a conveyor belt and exposed to photonic radiation.

From the radiological protection point of view, there are several differences in shielding design for facilities that use accelerators or high activity sources. Considering the irradiation phase and the rest phase (when there is no irradiation), a fundamental difference the ability to turn off the accelerator in the rest phase, while a high-activity source must be retracted in a vault, which generally is a pool located few meters below the irradiation room. The source rest position is chosen to be below the irradiation room due to safety precautions: in case of failure of the source handling mechanism (generally based on mechanical or air-compressed systems), this can still re-enter in the vault using gravity.

There are also differences in shielding barrier design for radiation produced by accelerators or high-activity sources. For accelerators, it is necessary to distinguish between primary and secondary radiation: primary radiation is that of the beam coming out from the accelerator, while secondary is made of all other radiations, essentially leak (coming out from the accelerator head shield) and scattering radiation, produced in scattering (similar to reflection) process on the walls and objects. As mentioned earlier, the secondary radiation is usually about two to three orders of magnitude less intense than the primary. Barriers are named as primary and secondary and designed depending on which radiation hits them. The accelerator beam is directed, for a determined amount of time, on the food/crop to be treated. This approach increases the load factor on the primary barrier in front of the accelerator and is directly stricken by the beam. Therefore, the primary barrier is considerably thicker than the secondary ones.

On the other hand, a high-activity source produces essentially an isotropic emission, so the dose is delivered almost equally on the entire conveyor belt pathway inside the hot room and on the walls. Therefore, in the case of symmetrical geometries, e.g. a squared room with the source in the center, all barriers possess essentially the same thickness and composition. In any case, to perform an accurate dose calculation on the walls, it is necessary to know in detail the characteristics of radiation field in the hot room.

Regarding the bunker entrance and exit for the conveyor belt, these are designed using a *wall maze* plan. This is necessary to reduce the radiations levels outside the hot room: radiation is degraded in energy and absorbed in the multiple scattering on the walls before being able to exit the maze (see Figure 6.1). Since the dose due to scattered radiation is orders of magnitude lower than that delivered by direct radiation, such design makes it possible to considerably lower the dose levels emerging from irradiation room without using any mechanical doors. These are generally very heavy (due to the large amount of material, especially lead), expensive, and subject to breakage caused by mechanical stress.

The energy of the gamma beam (E_γ <5 MeV) generally used for sterilization purposes is too low to induce neutron photo production in air (which is relevant only above 10 MeV) and consequently there is no argon

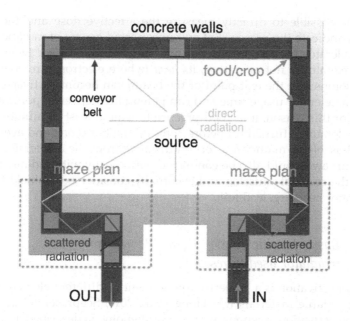

Figure 6.1 Sketch of irradiation room, with the maze plan designed for the entrance/exit of the conveyor belt.

activation problem (NCRP 2005a). ^{40}Ar naturally occurs in air and can be activated through neutron capture, becoming ^{41}Ar (β and γ emitter), which when inhaled causes internal exposure. This problem is typical in bunkers for medical accelerators (e.g. Linac's) with gamma energy greater than 7 MeV, and forced ventilation systems (with a determined ventilation rate) are used to lower the ^{41}Ar concentration in air before allowing it to enter the room. The ventilation rate is a compromise between this need and the necessity to not release in the environment too much radioactive gas, which may be left to decay within the room itself (NCRP 2005a).

In normal operations, workers are not exposed to radiation because facilities are highly automated, so all operations are remotely controlled from a room distant from the source. If a worker needs to enter the hot room, e.g. for handling the system or because of mechanical failures, a safety mechanism allows access only when the source is in the rest phase. To avoid any accidental exposure, every door to the hot room should be equipped with magnetic locks, so if these are inadvertently open during irradiation, the accelerator is immediately turned off or the radioactive source is retracted behind shielding or into the vault/pool.

For radiological protection, workers must wear personal dosimeters, Thermoluminescent Dosimeter (TLD) or electronic with direct reading, when entering the hot room. Electronic with direct reading in particular make it possible to directly measure the effective dose and therefore quickly indicate the presence of any unexpected source of radiation, e.g. the accelerator head or the source not completely shielded. Even when the accelerator is not operating, its *head* (where electrons are converted into photons and the exit point of the beam) can become activated after a long accelerator usage time and can potentially deliver important dose rates. For this reason, it is good practice to wait at least 10 minutes after the accelerator is turned off before entering the hot room, and even then it is advisable to maintain a certain distance from the accelerator head. Dose surveys should also be conducted periodically with radiometers to verify the residual emission and eventually consider extra shield for the accelerator head, e.g. lead foils.

6.3.2   Basics for shielding of ^{241}Am-^9Be neutron irradiator used in research

Neutron activation is a powerful tool to identify different elemental species in organic materials (De Hong et al. 2010, Seeprasert et al. 2017). Neutron activation reactions are nuclear phenomena described as

$$\,_Z^A X + n \to \,_Z^{A+1}X + \gamma_1 + \gamma_2 + \cdots$$

where a nucleus $\,_Z^A X$ absorbs a neutron and a new $\,_Z^{A+1}X$ nucleus is formed. Since the latter nucleus is an excited state, it loses its energy by emitting one or more gammas in a de-excitation process until reaching the stationary level, and eventually it decays again into another nucleus. Since the gamma energies are specific for each nucleus, by studying the energy of the emitted gammas it is possible to identify and quantify the $\,_Z^{A+1}X$ nucleus and therefore the $\,_Z^A X$ initial chemical species. Therefore, by irradiating a sample with neutrons and analyzing the emitted gamma with spectrometric techniques, the elemental composition of the sample can be studied.

If a nuclear reactor is not available, the most common neutron source used in research is the ^{241}Am-^9Be (Seeprasert et al. 2017), due to the good neutron flux (about 10^7 n/s over 4π for an activity of 37 GBq) and long half-life (432 years) (Holmes 1982). The average energy of the emitted neutrons is about 4 MeV, while the maximum energy exceeds 10 MeV. Therefore, these sources are well set to be used in activation analysis (Didi et al. 2017), under the condition to reduce the neutron energy below 0.1 eV in a thermalization process.

The Am-Be source emits neutrons according to the reaction:

$$\,_4^9 Be + \alpha \to \,_6^{12}C^* + n$$

An α particle is emitted by the ^{241}Am decay and it is captured by ^{9}Be, which is a light element with a neutron with low binding energy (1.7 MeV, because is uncoupled). After the neutron is emitted, the element is transformed in $^{12}_{6}C^{*}$, which is in an excited state. There is a prompt gamma emission (of about 4.5 MeV) due to the carbon nucleus de-excitation process, and this gamma emission must be considered in the facility shielding design, together with the neutrons emitted and their energy spectrum and other radiation fields surrounding the source, such as leptons and protons, as shown in Figure 6.2.

The irradiator presented allocates 6 Am-Be sources, placed symmetrically around a sample holder, and it is examined through simulations conducted with the Monte Carlo particle transport simulation code Particle and Heavy Ions Transport code System (PHITS 2020).

The neutrons emitted by (α,n) sources (but also those emitted in from spontaneous fission reaction as those of the ^{252}Cf sources) have high energies, well above the meV (1 meV = 0.001 eV) thermal range necessary for activation analysis. Since the probability of neutron absorption is considerably higher at low energies, a thermalization process is necessary to lower their energy before these can be actually used. This consists of multiple elastic scattering interactions inside the material, with the energy E_t *transferred in a single scattering event* given by

$$E_t = \frac{4A_{mass} \cdot E_n}{(A+1)^2}$$

where A_{mass} is the mass number of the nuclear target and E_n is the initial neutron kinetic energy.

It can be easily seen that the highest energy transfer happens for target material with $A_{mass} = 1$, which is why highly hydrogenated materials (e.g. polyethylene) are the most effective for the thermalization process (ideally the most effective is liquid hydrogen). Graphite or polyethylene layers are generally used for neutron thermalization, and it is necessary to consider carefully the secondary radiation fields (see Figure 6.3) arising in the neutron interaction within the material in the shielding design.

In the irradiator, the source array is surrounded by polyethylene and graphite layers, which provide both neutron thermalization and shielding. Observing the particle flux outside the polyethylene and graphite layers (shown in Figure 6.3), it is possible to see how important the contribution of high energy gamma is outside these layers, where neutron capture on hydrogen (2.2 MeV) and carbon (4.9 MeV) peaks are clearly visible. These photons also generate electron clouds, localized around the external surfaces of the irradiator, which are due to Compton scattering

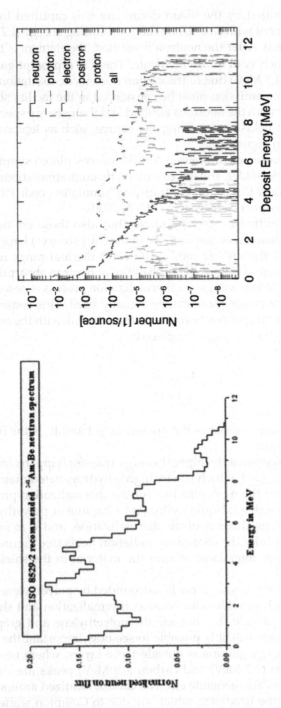

Figure 6.2 (*Left*) Energy spectrum of an Am-Be neutron source, simulated according to the recommendations of ISO 8529-2:2000 (ISO 2000). (*Right*) Flux of particles around the Am-Be neutron source in air. The interaction of neutrons with air gives rise to different radiation fields. The most important, after the neutron interaction, is the high-energy photon contribution due to the nuclear interactions of neutrons with air nuclei (N, O, Ar, C).

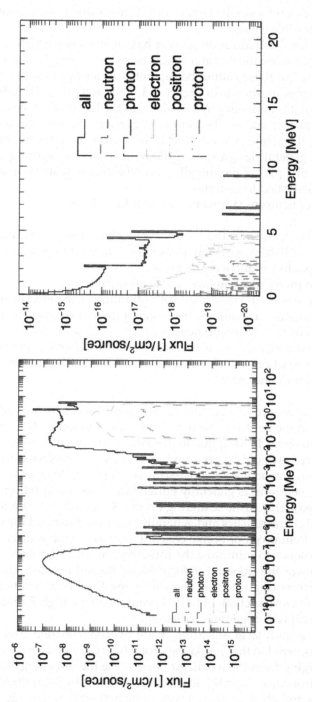

Figure 6.3 (*Left*) Thermalized neutron spectrum (red) obtained in the irradiator sample holder using 1-cm thick polyethylene layer and 10-cm thick graphite layer for thermalization analysis. (*Right*) Radiation fields outside the irradiator, shielded with 48-cm polyethylene and 10-cm graphite thick layers.

within the layers, and a small number of positrons due to pair creation, which is an accessible energy.

In general, the thermalization process has an efficiency which can be far from unitary values: the thermal neutrons can represent the main part of the spectrum, but these cannot be separated in terms of energy from the other neutrons, as shown in Figure 6.3. Therefore, in the irradiator shielding design it is necessary to consider also the high energy neutrons by using a multilayer and multi-compound barrier, which is also useful in minimizing the size of the neutron facility bunker, as shown in the following. The general design should follow this scheme: starting from the source, the first part of the shielding is dedicated to neutrons and the second one is dedicated to gammas.

A multilayer neutron-gamma barrier should include:

1. An iron layer, only in the case of very high energy neutron, i.e. greater than 10 MeV, as when accelerators are used to produce neutrons by photo-production processes.
2. A neutron thermalization layer made of concrete.
3. A neutron absorption layer made of cadmium or concrete. Given the cost of cadmium, it should be used only if there is not enough space for using concrete to absorb neutrons.
4. High-Z material, e.g. lead, is necessary to absorb photons originated in nuclear reactions and for absorption of thermalized neutrons in the cadmium/concrete layer.

The first three layers are for neutron shielding and the fourth for gamma. The iron layer in the neutron shielding is also used to lower the neutron energy down to about 10 MeV, due to inelastic scattering processes. Even if high-Z materials should be preferred, iron is usually preferred to lead because of its smaller activation cross-section. The concrete layer, used to thermalize the neutrons, is generally enriched with boron to reflect neutrons and increase the elastic scattering events. This kind of concrete has poor structural characteristics and generally is not used to build the walls of the bunker. Reinforced or borate concrete should be put in place with vibration techniques to minimize the inner content of air bubbles, since these would lower the shielding effectiveness. Regarding the cadmium layer, it is worth noting that this element is used not only for its large neutron capture cross-section, but also because, being a high-Z material, it has good absorption of gamma radiation.

Therefore, a multilayer shielding made of iron, concrete, and cadmium was considered in the simulations, for which the worst case is when neutrons emerging from the irradiator possess the highest energy (about 5 MeV). If the iron layer (1 cm thick) was followed only by 3-cm thick concrete, neutrons and photons are not well contained and can emerge from

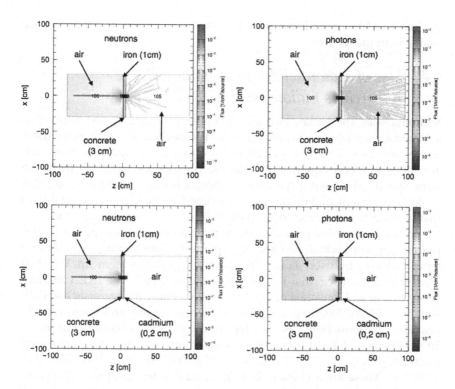

Figure 6.4 Upper left and right: Shielding design for neutron irradiator bunker, consisting of 1 cm thick iron layer, 3 cm thick concrete wall. Lower left and right: A 0.2 cm of cadmium layer is added to the previous configuration. The highest energy neutrons that can occur outside the irradiator (about 5 MeV) are simulated to impinge on the shielding in a beam configuration (worst case that can occur if there is some leak in the shielding). Neutrons and photons arising from interactions within the shielding are shown separately.

the irradiator (shown in Figure 6.4). Therefore, it is still necessary to add a 0.2-cm thick cadmium layer to adsorb neutrons and high energy photons originated by neutron interaction with shielding materials. This configuration provides good radiation containment, but the gammas emitted by neutron capture on cadmium (^{113}Cd, 12.23% of the natural cadmium) are still not shielded, and a lead layer (typically 3 mm) should be added as the final element of this multilayer and multi-compound shielding. In this way, the entire shielding is less than 5 cm thick and it is effective in shielding radiation and bringing the external dose at the required levels.

This simple case study has shown that the multilayer shielding approach is very effective in reducing neutron and gamma flux from the source and also the gamma flux generated by neutron interaction in the shielding material. This can be also achieved with an overall limited thickness, which constitutes an advantage when there is limited space available in the facility.

6.4   Use of radionuclides in agricultural research as radiotracers

Radionuclides are the key element of radiotracers, and as any radioactive substance, they pose a radiological risk. The key features that make them interesting for research in agricultural sciences, discussed in Section 6.2.1, also have as a consequence that the safety rules for handling them are quite specific. Following the correct approach in realizing the radiation protection system makes the risk associated to their use comparable to that of handling any other potentially hazardous chemical substance.

In this section we provide information on regulation, technical, and practical aspects, which can help to explain the functioning of safety systems and may be used as starting points for setting them up. We start from the risk source; that is, the radioisotopes used. The IAEA manual *Use of Isotope and Radiation Methods in Soil and Water Management and Crop Nutrition* (IAEA 2001) reports a list of the most used, and commercially available, radionuclides, as also reported by other authors (Alam et al. 2001, D'Sousa 2014). In increasing atomic number order these can listed as: ^{3}H, ^{14}C, ^{22}Na, ^{24}Na, ^{32}P, ^{35}S, ^{42}K, ^{45}Ca, ^{54}Mn, ^{55}Fe, ^{59}Fe, ^{60}Co, ^{64}Cu, ^{65}Zn, ^{86}Rb, and ^{90}Mo. These are all beta or beta/gamma emitters, and no alpha emitter is present and used as an unsealed source in experimental protocols. As discussed earlier, harm can be caused by external and internal exposure; therefore irradiation and incorporation, mostly but not only through inhalation, are the main radiological protection aspects to be considered. Other methods of introduction – ingestion, and absorption through skin or wounds – may still be considered, but these are usually prevented by using PPE and by forbidding eating and drinking in the area where radionuclides are manipulated.

Before going into detail on the radiation protection system approach, it is worth mentioning another important feature for handling radiochemicals as well as for non-radioactive chemicals: cross-contamination of the laboratory and its equipment can lead to false results. In addition to normal practices to avoid cross-contamination in place in chemical and biological laboratories, for radiochemicals interference with measurements can also occur without contact, e.g. from other sources nearby, possibly altering or invalidating the measurement results. It is therefore essential to consider the unwanted or accidental presence of other radiation sources and take the necessary steps to avoid it. However, since the impact of cross-contamination on measurements is outside the scope of the authors, this will not be given further consideration, except for those aspects directly involving hazards to workers as radioactive contamination.

6.4.1 Radionuclides used and radiotoxicity groups

The main characteristics, from the radiological protection point of view, for the radionuclides previously listed are reported in Table 6.2. These include the type and energy of the radiation emitted, the decay half-life, the specific activity, and the radiotoxicity group to which they belong.

Table 6.2 Radionuclide main characteristics and radiotoxicity group

RN	Emission type	Energy (MeV)	$T_{1/2}$[a]	As (Bq/g)[b]	Radiotoxicity group EURATOM (1984)	CEA Handbook (2002)	Research area(s)
H-3	Beta	0.018	12.46 y	$3.58\cdot10^{14}$	4	5	Pesticides, agrochemicals
C-14	Beta	0.155	5570 y	$1.66\cdot10^{11}$	3	4	Pesticides, agrochemicals
Na-22	Beta gamma	0.54 1.28	**2.58 d**	$2.31\cdot10^{14}$	3	3	
Na-24	Beta gamma	1.39 1.368, 2.754	**15.05 h**	$3.23\cdot10^{17}$	3	2	
P-32	Beta	1.7	**14.3 d**	$1.06\cdot10^{16}$	3	2	Pesticides, agrochemicals
S-35	Beta	0.168	**89 d**	$1.58\cdot10^{15}$	4	5	Pesticides, agrochemicals
K-42	Beta gamma	3.58 1.51	**12.47 h**	$2.23\cdot10^{17}$	3	3	
Ca-45	Beta	0.254	165 d	$6.60\cdot10^{14}$	3	4	
Mn-54	gamma	0.84	314 d	$2.87\cdot10^{14}$	3	3	Fertilizer label
Fe-55	Beta gamma	0.61×10^{-3} 5.9	2.94 y	$8.76\cdot10^{14}$	3	3	
Fe-59	Beta gamma	0.271 1.30	**44.3 d**	$1.84\cdot10^{15}$	3	3	Fertilizer label
Co-60	Beta gamma	0.31 1.17, 1.33	5.24 y	$4.18\cdot10^{13}$	2	2	
Cu-64	Beta gamma	0.57 1.34	**12.9 h**	$1.43\cdot10^{15}$	4	3	
Zn-65	Beta gamma	0.325 1.12	246 d	$3.05\cdot10^{14}$	3	3	Fertilizer label
Rb-86	Beta gamma	1.82 1.08	**18.68 d**	$3.01\cdot10^{15}$	3	2	
Mo-90	Beta gamma	1.085 0.122, 0.257, 0.511	**5.56 h**	$2.27\cdot10^{17}$	3	Not present	Fertilizer label

[a] Half-life v alues shorter than 100 days are reported in bold characters.
[b] Specific activity values have been intentionally rounded up to three digits. Accurate values are available from the LARA (LNHB 2020) and (NNDC 2020) database.

The radiation type and energy emitted are necessary for considering external exposure risks and appropriate shielding. The beta emission has a continuous spectrum, so the maximum energy is reported, while gammas have a discrete spectrum so the exact energies for emissions with highest yield values (yield values not reported) are provided. The half-life provides information on the time period a radioisotope is usable and also in relation to the waste management, discussed in Section 6.5. The specific activity is necessary to find the correspondence between the mass and activity of the radioemitters.

Finally, regarding radiotoxicity, this is defined as "the toxicity attributable to ionizing radiation emitted by an incorporated radionuclide and its daughters; radiotoxicity is related not only to the radioactive characteristics of the radionuclide but also to its chemical and physical state and to the metabolism of the element in the body or in the organ" (EURATOM 1980), and it is a way to summarize the risk associated with the radioisotope incorporation. Definition of the radiotoxicity groups takes into consideration the specific activity, the inhalation and ingestion annual limit of intake (ALI) (Section 6.4.4) in terms of activity and mass, and the exemption/clearance level (Section 6.5.1).

It is worth noting that different grouping approaches are present in the literature, as in *IAEA Technical Report Series No. 15 – A Basic Toxicity Classification of Radionuclides* (IAEA 1963), *International Commission on Radiological Protection – Report of Committee V on the Handling and Disposal of Radioactive Materials in Hospitals and Medical Research Establishments,* ICRP Publication 5 (ICRP 1964), the *European Council Directive 84/467/ EURATOM* (EURATOM 1984), the *Radionuclide Hazard Classification – The Basis and Development of a New List* by the National Health and Medical Research Council and the Standards Association of Australia (NHMRC 1993), and the *Radionuclide and Radiation Protection Data Handbook* by the French Commissariat à l'Energie Atomique (CEA) (Delacroix et al. 2002). Since the grouping approaches are slightly different, some radionuclides belong to different groups in different publications. Only the grouping proposed by EURATOM (EURATOM 1984) and CEA (Delacroix et al. 2002) are reported in Table 6.2: the first because is the basis for current European regulations, and the second because the publication provides quantities directly usable in designing and managing a radiation protection system. In this latter case, the risk groups are based on both external and internal exposure (Delacroix et al. 2002).

In both these classifications, the radiotoxicity increases from group 4 (or 5) to group 1. The specific nomenclature for Directive 84/467/ EURATOM (EURATOM 1984) and color code for CEA grouping (Delacroix et al. 2002) are reported in Table 6.3.

Because the risk is proportional to the amount of radioactivity held and manipulated, the radiological protection system becomes progressively

Table 6.3 Nomenclature and color code for radiotoxicity groups

Group number	EURATOM (1984)	CEA Handbook (2002)
1	Very high radiotoxicity	Red
2	High radiotoxicity	Orange
3	Moderate radiotoxicity	Yellow
4	Low radiotoxicity	Green
5	-	Blue

Note: Radiotoxicity decreases from 1 to 4 (or 5).

Source: Provided by EURATOM (EURATOM 1984) and CEA Handbook (Delacroix et al. 2002).

stronger with increasing radiotoxicity, activity manipulated, and operation performed. In a graded approach, safety features are progressively incorporated into the laboratory design (ICRP 1989). Therefore, in the *Recommendations of the International Commission on radiological protection – Publication 5* (ICRP 1964), three different laboratory types are considered: Level 1 (low level), Level 2 (intermediate), and Level 3 (high). Given the total radioisotope quantity needed, that is the maximum activity present at any time in the practice, and the radiotoxicity group of the radiotracers, the laboratory type necessary for handling them, is specified in Table 6.4. This information can be used both for existing laboratories, providing the maximum quantity that can be held, or when a new laboratory is set up, given the type of research foreseen and radionuclide to be held and their radiotoxicity; the correct laboratory type would ensure health protection and the requirement fulfillment.

The amounts in Table 6.4 are modified, according to the operation performed, by the multiplying factors listed Table 6.5.

This is because different operations imply different potential exposure and higher risks lead to lower multiplication factors. Clearly stocking radioactive source in storage is a rather inert situation and bears little risk, in opposition to grinding, where powder dispersion is potentially more

Table 6.4 Laboratory type based on radionuclides toxicity group and total activity held

Radiotoxicity group	Laboratory type		
	Type 1	Type 2	Type 3
1	< 370 kBq	370 kBq – 37 MBq	> 37 MBq
2	< 3.7 MBq	3.7 MBq – 370 MBq	> 370 MBq
3	< 37 MBq	37 MBq – 37 GBq	> 37 GBq
4	< 370 MBq	370 MBq – 370 GBq	> 370 GBq

Note: The radiotoxicity group is that provided by EURATOM (EURATOM 1984).

Table 6.5 Multiplication factors for activity levels depending on the operation
type performed

Operation type	Multiplication factor for activity levels
Simple storage	100
Very simple wet operation (e.g. preparation of aliquots of stock solution)	10
Normal chemical operation (e.g. analysis, simple chemical preparation)	1
Complex wet operation (e.g. manipulation of powder)	0.1
Working with volatile radioactive compounds	0.1
Dry and dusty operations (e.g. grinding)	0.1

hazardous. The values obtained by multiplying the radioactivity amounts
in Table 6.4 by the factors associated with the different operations provide
the correct laboratory type for the operation to be performed. However,
even if national regulations provide the exact criteria for compliance, and
even if these mostly follow international standards, it is worth noting that
these may use a different radiotoxicity evaluation, or different values for
allowed activity levels, multiplication factors, and laboratory type definitions from those reported in Tables 6.4 and 6.5.

6.4.2   *Laboratory design and requirements*

Each laboratory type entails specifications for the facility design, containment system, control equipment, and safety procedures, as well as
more in-depth training for the laboratory personnel. Further details are
available in publications such as ICRP Publication 57 (ICRP 1989) and the
IAEA Manual (IAEA 2001). The laboratory characteristics are specified in
national regulations and standards, mostly derived from international
regulations. For instance, in Italy the national standard is provided by
the Ente Nazionale Italiano di Unificazione [Italian National Unification
Institute] in UNI 10491:1995 (UNI 1995).

The general approach is to confine any contamination near the point
of use, using appropriate equipment and operating procedures to achieve
a radiological protection both within the facility and in the outside environment. The laboratory should be kept separated from other areas,
allowing entrance only through a so-called filter room, and categorizing
the laboratory working areas based on the amount of activity handled
and the corresponding potential dose for the workers (ICRP 1989). Other
general requirements are (ICRP 1989): allowing access to authorized personnel only, using cleanable and non-absorbing materials (for working
surfaces, floors, walls, and ceilings), ensuring the required air change rate

through a proper ventilation system, a containment system based on negative pressure (the high-risk areas are at lower pressure with respect to lower risk areas and the outside), ensuring the presence of hand-washing sink(s) (with controlled discharge into tanks), and emergency shower. Furthermore, operation with air contamination must take place in a fume cupboard (IAEA 2001) with negative pressure, the source must be stored into appropriate shielded cabin/room, and contamination monitoring equipment must be available (e.g. portable contaminameter and hand-foot monitors) at the classified area entrance/exit. The air released by the ventilation system, fume cupboard, and other equipment (e.g. digesters [IAEA 1973]) must be stripped of radiocontaminant by absolute filters, chosen according to the radionuclide and associated chemical form handled, and must comply with the national regulations. If operation with gamma emitters or intense beta sources are performed or foreseen, the need for shielding must be evaluated. If lead glass may be used, the weight and the space needed to allocate the shielding must be considered; in particular for workbench and fume cupboards.

Returning to the radioisotopes used in agricultural research (Table 6.2), the one with the highest radiotoxicity is ^{60}Co, belonging to group 2 (high toxicity). All other radioisotopes, including the commonly used ^{3}H, ^{14}C, ^{32}P, and ^{42}K, belong to group 3 or 4. The most common operations performed in agricultural research laboratories are preparation of aliquots from stock solution, transfer of aliquots to soil and plants, and drying of the experimental samples for analysis preparation, which have multiplication factors of 10 or 1. Therefore, since quantities normally manipulated are below 37 MBq, the laboratory type 1 is sufficient.

In the less common eventuality that larger activities are handled, e.g. a large amount of radioactive solution for soil-plant uptake experiments performed in large lysimeters, the laboratory type 2 is sufficient for total activity amount up to 37 GBq. In agricultural research, the use of larger radioisotope quantities is not known to the author, but in such instances, the information in Tables 6.2, 6.4, and 6.5 provide guidance for approaching national regulations.

6.4.3 Radiological protection operational guidelines and procedures

In addition to the laboratory design, the key features of the radiation protection system are operational procedures and good practice. In this section we provide some general information and guidelines for correct laboratory operations. The indications provided here should be considered as general guidelines, since an accurate evaluation of the radiological risks and prevention measurements requires specific information, and this is handled by the appointed RPE. The multiplication factors in

Table 6.5 provide some indication for the risks associated with the different types of operations.

In considering worker exposure, it is important to identify the risks and evaluate potential doses associated with the different experiment phases and operations to devise a proper mitigation strategy. For radiotracers, different phases are:

- Stock solution preparation
- Administration of radiotracers to plant and/or soil
- Experiment running
- Preparation of sample for measurement
- Experiment closure and management of the waste produced

For a correct evaluation, the characteristics of the sources involved and the operation details must be known. Radioactive waste management is discussed in Section 6.5, while the other phases are examined from the radiological protection point of view here.

Radiotracers are usually provided as soluble powder or in liquid form and then are diluted to obtain the stock solution, which may be further diluted to obtain the concentrations required for the experiments. Before opening the radiotracer containers, it is good practice to check them for integrity and identify any possible eventual leakage. In case of damage, possible contamination is investigated by measuring the contact dose and surface contamination with direct and indirect methods. Dose should be also measured (e.g. with a radiometer) once the radiotracer container is opened, and the result compared with expected values, as any large difference would point to possible anomalies.

Given the activity involved, stock solution preparation is normally one of the most dangerous operations. Initial dilution operations with a volatile or dusty compound, or operations that can produce volatilization, must be performed in a fume cupboard, and this is still recommendable for less volatile radiotracers. Usually stock solution is prepared at concentration from some tenth to hundreds of kBq/L, and then further diluted for use throughout the experiment, so operations are progressively less hazardous. However, since a degree of care and attention is required for all wet operations, the indications should be observed independently from the dilution factor.

When using gamma or intense beta sources, the need of shielding for the operators must be considered. Gamma sources lead or glass lead can be used, while for most beta sources plastic (e.g. PMMA) shielding is sufficient (Comar 1955). The principle of radiation attenuation and shielding were discussed in Section 6.3, and material attenuation factors (e.g. TVL for gamma and maximum range values for beta) can be found in NCRP publications (NCRP 2005a, NCRP 2005b, NCRP 2007), but the *Radionuclide and Radiation Protection Data Handbook* (Delacroix et al. 2002) also provides useful shielding TVL values for gamma and maximum range R^{max} values for beta for all the radionuclides in Table 6.2. For gamma emitters, the recommended attenuation factor is of about 1000 (ten half-value layers, or three tenth-value layers) (ICRP 1989), or at least of a factor high enough to reduce irradiation below the allowed limits (considering all the different workers' exposure). Beta radiation can be fully stopped by a thickness given by the ratio between maximum range and material density (see Section 6.3).

Transparent shields (lead glass for gamma and transparent plastic material for beta) are preferable to the use of mirrors, and if necessary these must be set so as to permit the use of long-handled tools or remote handling devices. The shielding must protect workers' entire bodies, the rest of the room, and adjacent areas if other workers are potentially exposed during operations. Consideration must be given to whether workbench and fume cupboards are strong enough to support the shielding weight.

PPE such as disposable rubber gloves (two pairs when handling solutions) and lab coats must be always worn, and in specific cases use of a dust mask or gas mask (with appropriate filtration/filter) and plastic apron can be considered.

Contamination risk, e.g. through accidental spillage, is another important feature to be considered. To reduce such risk, operations should be performed in easily cleanable, non-absorbing metallic or plastic trays lined with absorbent paper at the bottom of the tray and below it. Use of plastic foil to protect working surface or equipment is also a good practice. The same approach should be taken with all operations involving handling of radioactive solutions or objects that contain it, e.g. when loading devices or preparing plant or soil samples for measurement. Once the operations have been completed, devices and equipment must be checked for contamination using portable contamination monitors and eventually performing wipe tests. PPE contamination should also be checked using the same instrumentation or the hand-foot monitor (and its detachable probe).

Radioactive solution is normally administrated to soil for nutrient and fertilizer experiments, and to plant leaves and/or soil for pesticide and herbicide experiments. This can be done using syringes, micro-syringes, pipettes, bottles, or other laboratory equipment. In the case of open field experiments, where sprayers may be used, specific protection measures must be considered (Milburn et al. 1959). All the equipment used should be in principle considered as radioactively contaminated, so it must be properly washed/decontaminated or disposed of as radiological waste. Particular attention must be paid to contaminations of plastic when using compounds/chemical with a high organic affinity.

Indoor experiments are usually performed in pot or soil columns (Ashworth et al. 2003) located in the laboratory itself or in a greenhouse. The dedicated space should be separated from its surroundings, and access must still be controlled. When using volatile compounds/radiotracers (e.g. ^{14}C, ^{35}S) controlled atmosphere rooms are worth considering. If high activity values are administered, the environmental dose should also be monitored by using direct reading or TLD dosimeters positioned at 1 m from the experimental devices.

For outdoor experiments performed in open fields (Milburn et al. 1959) or lysimeters (Wheater et al. 2007), it is necessary to have a dedicated set of protective measurements. The field or lysimeters must be located away from buildings and other human placement, such as farms and water supply. The area should be delimited with fences, and access must be controlled to avoid unauthorized entrance. Special care must be taken to avoid or minimize any leakage (e.g. due to flooding following intense rain) to surrounding areas, and in particular to aquifers. The use of low-pressure sprayers and nozzles producing large droplets are highly recommended; spraying should be avoided in windy conditions (wind speed > 10 km/h), and spray equipment should be covered with plastic cling film or polythene sheeting. The area should be monitored for contamination through air sampling and collecting samples from the area borders and nearby areas.

Preparation of samples for measurement should be performed using the same precautions discussed earlier. In this case, the activity content is usually lower than in the previous operation, but contamination of the tools and equipment used (e.g. sieves, scales, glassware, centrifuges, ovens, magnetic stirrers, digesters) should be monitored for contamination. Any operation potentially leading to volatilization should be performed in a fume cupboard. Tools can be decontaminated using proper methods and reused after decontamination has been verified (ISO 1988). Glassware can be washed with monohydrated citric acid or potassium hydroxide solution before rinsing it with distilled water. All material used for decontamination, either solid or liquid, must be considered as radiological waste.

The experiment closure implies the removal of potentially contaminated material, which should be tested for dose rate and superficial contamination (direct and indirect methods) before handling. The contaminated and radioactive material that will not be reused is considered radiological waste, as discussed in Section 6.5.

In general terms, when planning a set of experiments, in order to fulfill the ALARA principle, it is recommended to choose radiotracers that allow performance of experiments with a minimal level of radioactivity and radiotoxicity (IAEA 2001). Performing a pilot experiment is a sensitive approach for establishing such minimum values, before running it at full scale. Moreover, this allows researchers and operators to become familiar with the different procedures and to fine-tune the experimental protocol considering both the achievement of experimental results and radiological protection issues. It is always a good approach to first perform any delicate and potentially hazardous operations with mock solutions, so to acquire familiarity with each step and preventively identify criticalities and unforeseen risks.

There are also a few general good behavior rules which, even if rather simple, are important in reducing the risks. First, it is preferable to have only the activity amount needed present in the working area, and to store the remaining in fume cupboards or source cabinets. All risks can be reduced by simple good housekeeping and working habits – that is, performing the operation carefully, with full attention (Comar 1955). It is important to always keep in mind the radioprotection rules and procedures: these must be written so not to overwhelm or to needlessly restrict the operators' work (Comar 1955). Providing appropriate training is very important, and operators must gain hands-on experience in steps, starting with low-risk operations and radiotracers, so they will be able to correct themselves from inadequate behavior. There is no substitute for experience.

As noted earlier, in designing the radiation protection system, the radioactive sources (e.g. radionuclides) and the operations carried out must be carefully considered, as each operation provides a particular dose to operators. This allows identification of the highest dose contribution and sets the pertinent reduction or risk mitigation actions. By summing up the type and number of operations performed, the annual dose can be estimated in advance.

In addition to information on the source, the dose evaluation is carried out using parameters and quantities that can be taken from international publications and literature. Many quantities of practical use are provided in the *Radionuclide and Radiation Protection Data Handbook* (Delacroix et al. 2002), such as exposure rates (mSv/h) for beta and gamma radiation from different source geometries (point source, infinite plane, 10-mL glass vial, 50-mL glass beaker, and 5-mL syringe), normalized to 1 MBq or 1 MBq/m^2 as appropriate, and shielding thickness for both beta (total absorption) and gamma (HVL and HTV) radiation, and the ALI and derived surface contamination limit (DSCL), discussed in the next section, the dose for hand contamination skin dose (droplet and uniform), and volatility factors. Regulatory quantities such as exemption values or committed dose coefficients are also provided.

6.4.4   Environmental monitoring program

Environmental monitoring is a key feature to ensure both worker safety and correct performance of operations. This consists of irradiation dose and contamination measurements, performed with an established frequency and following specific situations (e.g. particular operation, suspected contamination). It is a way to confirm or forecast dose values, identify abnormal situations, and support the fulfillment of regulation requirements.

For monitoring areas where operations are routinely performed, badge X-gamma dosimeters (TLD) or fixed radiation monitors are normally used. While the first device measures the cumulative dose throughout the exposition period, and to obtain the readout it must be sent to the dosimetry center for reading, the second device provides direct reading in real time. In both cases, the devices should be positioned sensitively, i.e. comparably with the worker's position during operation. The recorded value can provide direct checks as well as being useful for later dose evaluation. For specific operations, direct dose reading (in mSv/h or μSv/h) is provided by portable dose monitors as radiometers (see Section 6.4.6). The readout values must be compared with the natural environmental dose background, usually in the range of 0.1 to 0.2 μSv/h, or the value measured away from the radiation source, as the background value could be higher due to radiation from construction material (usually about 0.3–0.4 μSv/h).

For surface contamination, direct and indirect methods are used (ISO 2016a, ISO 2016b). Direct measurements use contamination monitors which provide the total contamination and are able to quickly identify contaminated spots. Portable monitors provide readout values in *cps* (*counts per second*) for alpha and beta radiation, and the correspondence to Bq/cm^2 is provided by the instrument-specific calibration factor (Section 6.4.6). The readout values must be compared with the environmental background and previously established investigation and alarm levels. Care must be used in interpreting the results when measurements are performed on surfaces that are not flat, or absorbing material, especially for alpha values. In any case, this type of instrument has a lower sensitivity compared to the wipe test, which by "wiping" the surface/objects with paper filters is able to remove a contamination fraction, providing the removable contamination. The wiped surface is usually equal to 100 cm² for a small area or object, or 300 cm² for a larger surface or object, and the wipe position must be recorded for later evaluations. The filters are then read in a total alpha/beta counter to obtain the superficial contamination in Bq/cm^2 which, due to a higher detection efficiency, normally provides a higher sensitivity to the method, The total contamination C_{tot} (Bq) can be obtained from the removable contamination, C_{rem} (Bq/cm²) by knowing the sampled area A(cm²) and the removal factor (F_{rem}), usually equal to 0.1 (10%):

$$C_{tot}(Bq) = C_{rem}\left(\frac{Bq}{cm^2}\right) \cdot \frac{1}{F_{rem}} \cdot A(cm^2) \qquad (4.1)$$

In routine contamination controls, working surfaces and floors as well as fume cupboards and laboratory equipment (e.g. tools, gauges, glassware, ovens, digesters) normally used should be controlled. In controls the investigation can be extended to less-used equipment, including door handles, personnel PPE, lockers, laboratory notebooks, computer keyboards, and any object that potentially can have been contaminated.

Air sampling is also commonly used to assess the airborne contamination, i.e. in the atmospheric particulate. This is done by pumping air through a paper filter, noting the air volume sampled, and measuring the filter on a total alpha/beta counter. In particular, this operation is necessary in the case of extensive use of volatile radioisotopes (e.g. radio iodine).

Investigation and alarm values are usually set in correspondence to 1/3 of annual dose limit and the entire annual dose limit, which is obtained by considering 2000 working hours per year.

In the case of external irradiation for Cat. A–exposed workers/supervised areas with a 20 mSv/y limit, these correspond to an hourly dose rate of 3 and 10 µSv/h, respectively. If the exposure limit is 6 mSv (Cat. B–exposed workers/controlled areas), the previous values are multiplied by 6/20 = 0.3, providing about 1 and 3 µSv/h, respectively.

Ti establish alarm and investigation levels for internal contamination, it is useful to refer to the *annual limit of intake (ALI)* values, even if these are not regulatory quantities. The ALI for each radionuclide and intake route (e.g. inhalation) correspond to the activity taken up by the worker which would reach the 20-mSv committed dose value. The ALI value can be expressed in activity (Bq) (EURATOM 1984), or mass (g).

For surface contamination, several features must be considered: transfer from contaminated surface to the atmosphere and consequent inhalation, ingestion, transfer to the skin leading to external dose to extremities, and whole-body exposure to surface contamination. A non-regulatory quantity makes it possible to consider all these features, the *derived surface contamination limit (DSCL)* (Delacroix et al. 2002). In calculating the DSCL values, the ALI values are used, therefore also DSCL corresponds to 20 mSv/y and 500 mSv/y for the whole-body effective dose (external and committed) and skin irradiation equivalent dose, respectively. If the exposure limits are those for Cat. B workers (6 mSv/y and 150 mSv/y, respectively), the DSCL values should be multiplied by 0.3.

Regarding the activity concentration in atmospheric particulate, the *derived air contamination (DAC)* is often used (IAEA 1999), even if this quantity is also a non-regulatory one:

$$DAC\left(Bq/m^3\right) = \frac{ALI_{inh}(Bq)}{R(\frac{m^3}{h}) \cdot T(h)}$$

where R(m³/h) is the adult respiration rate, equal to 1.2 m³/h, and T_{work}(h) is the number of hours worked per year, normally assumed equal to 2000 h/y. The air concentration equal to the DAC(Bq/m³) value implies that the 20-mSv annual limit is reached. If the dose limit is 6 mSv (Cat. B worker), once more the value obtained must be multiplied by 0.3.

If more than one radionuclide is present, there are two approaches. The first precautionary approach is to consider the most dangerous radionuclide and assume that all activity is due to this, and therefore consider only the DAC and DSCL for it. A more realistic approach, but not always feasible, is to use the DAC_i and $DSCL_i$ values for each ith radionuclide, and also measure the airborne and superficial contamination for each one and compare these values to the respective limit, according to the follow equations:

$$\sum_i \frac{C_i(Bq/m^3)}{DAC_i(Bq/m^3)} \le 1$$

$$\sum_i \frac{C_i\left(\dfrac{Bq}{cm^2}\right)}{DSCL_i\left(\dfrac{Bq}{cm^2}\right)} \le 1$$

where the sum is over the radionuclide considered. This method is more time consuming, especially concerning the measurements, and requires obtaining the concentration for each radionuclide by spectrometric analysis or other methods, which are not always practically feasible and easy to handle.

6.4.5   *Worker monitoring program*

In regard to workers' physical surveillance, they must be monitored for external irradiation, radionuclide incorporation, and contamination. The establishment of a monitoring program is necessary, and type of measurement, frequency, and other features depend on the radionuclides used and the estimate dose associated with the different operations. The IAEA provides detailed information on these aspects (IAEA 1999, IAEA 2018a).

For external irradiation, workers are normally given a standard badge X-gamma dosimeter (TLD) for whole-body dose and X-gamma-hard beta ring/bracelet dosimeters for doses to the extremities. For soft-beta radiation (low-energy beta), specific thin dosimeters for the extremities are also available. The dosimeters are read at regular intervals (e.g. 45 or 90 days, or according to RPE responsibilities) and provide the cumulative dose for the period. However, for specific operations when the dose must be known immediately, use of an electronic dosimeter which provides immediate readout (integrated over the exposition time) is recommended.

For internal contamination, a monitoring program must be established based on the specific risk. Two types of measurement are available: in vivo, i.e. through high/low energy body counter, and in vitro, i.e. analysis of radioemitters in bioassay (e.g. urine and feces). Both approaches are usually used in combination to provide the activity incorporated, which is then used for evaluating the effective committed dose (ICRP 1997). In vitro measurements for beta and gamma emitters include the periodic sampling and analysis of urine excretion, which can provide the total beta and gamma concentration through gross counting, and gamma emitter can also be quantified through gamma spectrometry. In vivo body counting techniques, using a high purity germanium (HPGe) detector, are able to provide both identification and quantification of incorporated gamma emitters. The method is applicable to the whole body or a portion of it (e.g. torso, thyroids) and even if generally less sensitive than bioassay measurement, it has the advantage of providing measurement results in a much shorter time.

The need for monitoring incorporation of specific radionuclides is established by the effective risk of radionuclide incorporation. In general, if it is unlikely that annual committed effective dose would exceed 1 mSv, the individual monitoring program may be unnecessary, but the workplace monitoring should be still undertaken (IAEA 1999).

The opportunity to monitor a specific radionuclide can be evaluated on the basis of the decision factor D, given by the relation (IAEA 1999):

$$D = \frac{d_{es}(Sv)}{d_{ref}(Sv)}$$

where d_{ref} is the *reference dose* (IAEA indicates 1 mSv), and d_{es} is the *estimated annual dose*, which is calculated as

$$d_{es} = f_{fs} \cdot f_{hs} \cdot f_{ps} \cdot e_{inh}(Sv/Bq) \cdot A(Bq)$$

where f_{fs} is a *factor accounting for the radionuclide physicochemical properties*, set normally to 0.01 (or 0.001), f_{hs} is a *factor accounting for the operation performed* (ranging from 0.01 for stock solution storage to 100 for dusty and dry operations), f_{ps} is a *factor for the protective laboratory equipment used* (1 for bench, 0.1 for fume hoods, and 0.01 for glove box operations), e_{ihn} (Sv/Bq) is the *committed dose factor* (the value for AMAD 5 μm is usually taken) and A is the total radionuclide activity in the workplace. For different operations and radionuclides, all the estimated doses must be summed up and compared to the reference dose to establish the individual monitoring program.

The monitoring frequency is established depending on the analysis method sensitivity and associated uncertainty, the annual dose limit, and the potential dose for the operation performed. It is also important to establish a point zero incorporation for the workers before they begin working with radionuclides or the assigned activities are modified. It is therefore necessary for the laboratory manager to promptly communicate to the RPE any foreseen change of the workers' activities, so internal contamination protocol can be varied accordingly and eventually point-zero value taken.

Workers' external contamination is evaluated with direct methods when they exit the classified areas with a hand-foot monitor, and with a portable contamination monitor following operations carrying potential risk or in case of an incident (e.g. spillage). In the latter case, the head-to-toe technique is used: that is, starting with the head and moving the detector slowly to the arms, palms of hands, and back, torso, legs, and feet.

Furthermore, in case of accident or suspected contamination, the measurement of nasal mucus can provide information on eventual inhalation. Even if this is a screening method, not providing good quantification for the intake, it is important for establishing whether inhalation has taken place, and if so, the time it occurred, as this last information is pivotal for evaluate the committed dose. The actual intake can still be obtained by bioassay measurements.

6.4.6   Measurement equipment

The measurement equipment that should be present in the laboratory has been noted in the previous section. This essentially is composed by:

- A radiometer; e.g. a Geiger-Muller or ionization chamber, calibrated in term of dose equivalent, for measuring dose rate
- A contaminameter, or portable contamination monitor, which is sensible to beta/gamma and alpha radiation, used to monitor surface contamination
- A hand-foot monitor, for measuring worker contamination
- Paper filter for executing a wipe-test, and a total beta/alpha counter used to quantify the contamination on the filter

All equipment must undergo calibration for the quantity measured, and the calibration must be repeated periodically, generally every 3 to 5 years (or depending on the manufacturer's recommendation), by a certified institute/laboratory. Good functioning and metrological confirmation tests are performed more frequently and can be carried out by the laboratory personnel under directions provided by the RPE. The workers should also be instructed on instrumentation use and data evaluation. Detailed descriptions of instrumentation for radiation measurement are provided by many authors, among which Knoll (2010) deserves mention for accuracy and completeness.

6.5   Radioactive waste management and disposal

A laboratory using radiotracers can produce some amount of radioactive waste originated by using unsealed sources which can also contaminate laboratory items and the environment. In general, the radioactive waste amounts are rather small and can be safely managed within the laboratory practice before disposal or conferment to specialized companies. It is important to consider the waste management strategy from the beginning, as it may more costly to include at a later stage.

As any radioactive source, radioactive waste must also fulfill national regulation requirements, which, again, follow international guidelines and basic safety standards. A short review of regulations is provided in the section 6.5.1, with the criteria for exemption and clearance, and the direct consequences of the waste management strategy.

In the section 6.5.2, the practical guidelines provided by the IAEA for radioactive waste handling and management are presented, together with some practical advice on the subject. However, exact indications and strategies depend on the specific radionuclide and facility features and national regulations and are usually provided by the appointed facility RPE.

6.5.1   Regulation overview and exemption and clearance criteria

EURATOM defines radioactive waste as follows: "radioactive waste means radioactive material in gaseous, liquid or solid form for which no further use is foreseen ... and which is regulated as radioactive waste by a competent regulatory authority under the legislative and regulatory framework of the Member State" (EURATOM 2013).

The prime responsibility of the waste rests with the license holder, so the practice is responsible for its part in the general waste management (EURATOM 2011). This includes handling and temporary storage at the practice, disposal according to regulations including conferring to a specialized company (EURATOM 2011), and the characterization and classification in accordance with requirements (IAEA 2009a)

As discussed in Section 6.2.3, practices can be in the exemption, notification, or authorization regime. A special case is that of practices in the exemption regime, as also the waste is automatically in exemption and it can be disposed as regular waste without any further evaluation (IAEA 2004). On the other hand, for practices in the notification or authorization regime, the radioactive waste management strategy is provided during the initial notification/authorization process, and the radiological waste must undergo an accurate evaluation and management process before being disposed of. Arrangements are made on behalf of the authorization holder, by the RPE, with the aim of not having any unnecessary exposure of the workers and no radiological relevance to the population. Given the amount of radioactivity normally involved for agricultural research, such laboratories can follow the guidelines and indications discussed by the IAEA for radioactive waste produced by the use of radionuclides in medicine (IAEA 1998, IAEA 2000).

Waste is managed and disposed of according to its category, and categories are fixed by national regulations based on potential harm, which in turn depends on the radiotoxicity and half-life of the radionuclides contained, their concentration and total activity, the waste physics-chemical form, and its heat production. The IAEA provides six categories for nuclear/radiological waste, from very low to high risk (IAEA 2009b), which normally constitute the basis for the categories defined in national regulations. Leaving intermediate and high level waste (ILW and HLW) out of the discussion, only the first four categories are of concern for agricultural research laboratories, since most of the waste from such installations will fall into the low-risk categories 1 to 3, and possibly sometime into the 4 category (e.g. for large installation/facilities). These are listed as follows (IAEA 2009b):

1. **Exempted Waste (EW):** Waste that meets the criteria for clearance, exemption or exclusion from regulatory control for radiation protection purposes.
2. **Very Short Lived Waste (VSLW):** Waste that can be stored for decay over a limited period of up to a few years and subsequently cleared from regulatory control according to arrangements approved by the regulatory body, for uncontrolled disposal, use or discharge. This class includes waste containing primarily radionuclides with very short half-lives often used for research and medical purposes.
3. **Very Low Level Waste (VLLW):** Waste that does not necessarily meet the criteria of EW, but that does not need a high level of containment and isolation and, therefore, is suitable for disposal in near surface landfill type facilities with limited regulatory control. Such landfill type facilities may also contain other hazardous waste. Typical waste in this class includes soil and rubble with low levels of activity concentration. Concentrations of longer-lived radionuclides in VLLW are generally very limited.
4. **Low Level Waste (LLW):** Waste that is above clearance levels, but with limited amounts of long-lived radionuclides. Such waste requires robust isolation and containment for periods of up to a few hundred years and is suitable for disposal in engineered near surface

facilities. This class covers a very broad range of waste. LLW may include short lived radionuclides at higher levels of activity concentration, and also long-lived radionuclides, but only at relatively low levels of activity concentration.

As from the category definition, waste belonging to the EW, VSLW categories can be then disposed of as regular waste eventually after clearance limits have been reached through radioactive decay, while waste belonging to VLLW or LLW is conferred to specialized and authorized companies which will be then responsible for its management strategy, including mid- or long-term storage and final disposal. In any case, waste management can be costly, and therefore it is necessary to account for strategies that would reduce the waste production and provide for safe short-term storage.

The overall approach to waste given by IAEA is summarized in Figure 6.5. The first consideration is if the practice is in exemption regime. If so, the practice is already outside the regulatory control, and the same for the waste, which can be disposed of as regular (non-radiological) waste. Otherwise, if the practice is in the regulatory regime (notification or authorization), it is necessary to evaluate if the waste meets the clearance criteria. Clearance means that the waste, which is initially within the regulatory framework, and can be excluded from regulatory control because it is not radiologically relevant. The criteria for non-radiological relevance were discussed in Section 6.2.3, and the IAEA has elaborated derived activity and activity concentration limits which ensure the respect of the criteria for the disposal scenario (IAEA 2004, EURATOM 2013, IAEA 2014).

There are two types of clearance-derived limits (IAEA 2004, EURATOM 2013, IAEA 2014): the general clearance limits provided for each radionuclide in term of activity concentration and applicable to any amount or radioactive waste (IAEA 2004, EU 2000), and more relaxed limits for moderate amounts (in the order of 1 ton) provided by the activity and activity concentration exemption values. In this latter case, one of two limits makes it possible to fulfill the clearance criteria.

In cases where the waste is above the clearance levels and contains short-living radionuclides, it shall be evaluated if clearance can be met by storing it and letting it decay in order to reach limit values in a reasonable time (a few years). Normally the decay time should be of 10 half-life, but it depends on the initial concentration values. In any case, it must be remembered that intentional or deliberate dilution in order to fulfill the concentration limit is strictly forbidden (EURATOM 2013, IAEA 2014).

Meeting the clearance limit still requires that the disposal must always be authorized by the regulatory body, and disposal can take place only after the authorization has been granted and the proposed disposal route

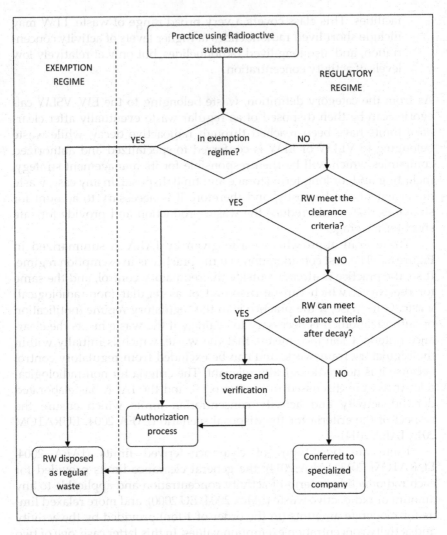

Figure 6.5 Diagram for radioactive waste management.

approved. The practice must submit a request that supports and justifies the clearance through a radiological impact assessment, and also contain waste characterization measurements, conducted according to specific requirements (IAEA 2004). If the approval is denied, or the clearance criteria are not meet even after decay, the waste must be conferred to a specialized and authorized company which will be in charge of the waste management, including the short- or long-term disposal. In addition, there are cases when the waste does not fulfill the clearance limits, but the radiological non-relevance can be proved through specific assessment.

These cases are handled on an individual basis and according to national regulation. If clearance is still granted, the disposal route is determined by that used in the assessment proving the radiological non-relevance.

It must be noted that the general clearance-derived limits given by IAEA (IAEA 2004, IAEA 2014) are valid only for solid and not for gaseous/airborne or liquid effluent. However, for moderate amounts such as those treated by research laboratories, these are provided in *Clearance of Materials Resulting From the Use of Radionuclides in Medicine, Industry and Research* (IAEA 1998).

As the diagram in Figure 6.5 was drawn following international regulations, there may be differences in national regulations, so this is only indicative of the actual approach to waste management strategy. For instance, in Italy there is a limit (T_{lim}) on the radionuclide half-life of 75 days which separates short- and long-living radioactive waste, and a concentration limit of 1 Bq/g for radiological relevance. Without going into detail, for short-living radionuclides the clearance can be granted only if the concentration is below the 1 Bq/g OR the total activity is under the exemption value (A_{ex}) stated in the national regulation AND below the clearance level (CL). For long living ones, the concentration must be below the clearance level AND below 1 Bq/g OR the activity below exemption limit. These conditions can also be summarized in Boolean algebra as below:

- Short-living ($T_{1/2} < T_{lim}$): C < 1 Bq/g or ((A < A^{ex}) and (C < CL))
- Long-living: ($T_{1/2} \geq T_{lim}$): C < CL and ((C < 1 Bq/g) or (A < A^{ex}))

Regarding the clearance level, for mixtures of several radionuclides the following equation is used to establish whether the waste fulfills the clearance criteria:

$$\sum_i \frac{C_i\left(\dfrac{Bq}{g}\right)}{CL_i\left(\dfrac{Bq}{g}\right)} < 1$$

where C_i (Bq/g) is the mass concentration of the ith radionuclide, and CL_i (Bq/g) is the corresponding clearance level values. For some materials and disposal routes, the clearance level can also be provided in terms of superficial concentration (Bq/cm^2), and this must be evaluated in terms of items/material and compared to the clearance limits (IAEA 2004).

The ability to reach clearance concentration by radioactive decay has two immediate consequences on waste management strategy. The first is that, compatibly with the experimental aims, it is always preferable to use short-term radionuclides. In this manner, the resulting waste can be then disposed after a limited storage time as regular waste, which is less costly. It is therefore necessary to plan a storage room (with appropriate shielding) in the facility, and to consider in the radiation protection system the waste control and management, and the measurements that will be performed to verify that clearance levels are achieved. The second is that, in order to have a cost-effective waste management strategy, the short-living and long-living radionuclides must be kept separated. If a waste item is contaminated with radionuclides of both types, it will follow the disposal route of the longer-living ones. In addition, even if items are contaminated by two short-living radionuclides, the storage time to reach the clearance level is determined by the one with the longer half-life.

6.5.2   *Waste management practical guidelines*

The approach to waste management for medical facilities is discussed in detail by the IAEA (IAEA 1998, IAEA 2000), but the same approach is still valid for radiotracer laboratories since the amounts of radioactivity used in them are similar. In general terms, the waste produced by a radiotracer laboratory is related to the following operations:

- Stock radioactive liquid solutions
- Plant/soil to which radioactive tracers were administrated
- Chemical processing of experimental specimens and samples
- Liquid waste arising from glassware washing/rinsing and liquid scintillation
- Disposable experimental/laboratory items/devices (plastic container, trays, pipettes, etc.)
- Single-use PPE.

From the radiological protection point of view, the same principles discussed earlier for radioactive sources also apply to safety and management of radioactive waste. The guidelines are provided by IAEA (IAEA 2000) and summarized here:

- Minimize the generation of radioactive waste, in terms of both activity and volume, using appropriate design measures, facility operation, selection, and control of materials, and the implementation of appropriate procedures.
- Waste volume minimization can be achieved by using available treatment technology and processes such as compaction, incineration, filtration, and evaporation.
- Segregation of different types of materials is used to reduce radioactive waste volume and facilitate its management.
- Minimizing the spread of radioactive contamination, as it leads to the production of radioactive waste, by maximizing containment efforts and minimizing the secondary waste creation.

- Separate valuable materials from waste and clear valuable materials for recycling and reuse.

In particular, practical implementation of waste minimization can be achieved by minimizing (IAEA 2000):

- Use short-lived radionuclides whenever possible, which can be decayed prior to disposal, to minimize waste activity.
- The waste volume is reduced by not taking non-essential non-radioactive materials into controlled areas, as this reduces potential cross contamination and the need for decontamination or disposal.
- The most effective step is to reduce waste at the source, so proper experiment design is a key step, as it may reduce waste production by several orders of magnitude.

The following general features should also be considered in relation to facilities generating radioactive waste (IAEA 2000):

- For radioactive and biohazardous waste, use the most effective and reliable technology, taking into account the safety requirements.
- To achieve segregation of radioactive materials from non-radioactive materials the latter should be kept out of classified areas.
- For each application, the minimum quantity of radioactive materials should be used.
- Radioactive material containment and packaging should be adequate for retaining the contents without resulting in unnecessary volume.
- When persons or items leave classified areas, decontamination should be performed if necessary, and adequate control to avoid cross-contamination or excessive production of radioactive waste should be maintained.
- Experiments should involve the minimal amounts to for obtaining scientifically valid data.
- Spent sealed sources should be returned to the vendor, using the original packaging material when practicable, whenever possible.
- Radioactive waste must not be mixed with other materials, especially chemicals, hazardous materials, biohazardous materials, and other regular waste.

In the laboratory, waste must be kept separated from other items and confined. Gaseous waste, due to its nature, is kept within the laboratory environment by the de-pressurization system. Discharge in the environment is subject to authorization and within prescribed amounts and concentrations (IAEA 1998). In all other cases, it is pumped by the ventilation system onto absolute filters, which can retain up to 99% of the airborne activity. The same is also valid for when produced in fume hoods. The filters are then treated as solid waste (IAEA 1998). Liquid waste, arising for instance from the washing of glassware, laboratory sinks, unused radioactive stock solution, chemical processes on experimental specimens, or used scintillation organic liquid, must be collected and stored in dedicated tanks, and any leakage must also be contained and confined. Normally these tanks are built into the facility design. The solid waste is easy to manage, as this can be placed and stored in metallic canisters.

At the laboratory facility it is then necessary to allocate for a dedicated deposit for waste storage. Its characteristics must be carefully evaluated based on the amount and type of waste produced. The deposit must have limited and controlled access, and all waste must be appropriately packaged, and the packages labeled according to regulations (IAEA 2006, IAEA 2008, IAEA 2018b). Waste records must also be maintained at the facility.

The exposure generated by the waste must be controlled using radiometers (e.g. GM counters or ionization chambers calibrated in terms of effective dose), as well as waste containers for gross alpha/beta superficial contamination, through portable contamination monitors or wipe-tests. Spectrometric methods can also be used to identify and quantify alpha and gamma emitters.

Before placing the waste into canisters, it is good practice to place it into sealed plastic reinforced bags. This is particularly relevant for experimental specimens or any chemically active material. Sharp objects (e.g. needles, syringes, pipette tips, blades etc.) must be placed in appropriate containers.

Glassware can be washed and reused if provisions are made to collect the liquid. It is good practice to wash it with monohydrated citric acid or potassium hydroxide solution before rinsing with distilled water.

These practical advice points complete the overview of the issues related to radioactive waste management. However, as noted, the exact waste management strategy is decided upon by the RPE of the facility, accounting for the type of radionuclide used, the experimental procedures, any specific issue at the facility, and the national regulations to be fulfilled.

6.6   Concluding remarks

The main features concerning the design and implementation of a radiation protection system have been presented and discussed, as well as practical guidelines for many different situations related to the use of radiation in high-intensity irradiation facilities, the use of radiotracers in agricultural research laboratories, and radioactive waste management.

The starting point is always an accurate risk evaluation; precise knowledge of the radioactive sources is necessary, and the risk is then managed and kept under control through a series of actions, procedures, equipment, and so forth, which are provided in international and national publications. Although these are very important, there is no substitute for experience; therefore the practice and its operational aspects must be known in depth. It is always a good practice to approach the issues in a very precautionary manner, eventually relaxing the approach once more experience and data on specific features are gained. On the other hand, even though safety is the priority, a sensible radiation protection system should not be too invasive and should allow the facility to carry out the operations without useless burdens and complications. In this regard, collaboration and mutual understanding among the different actors, as the license holder, workers, laboratory managers, RPE, etc. is essential and can ensure the correct approach and application for the radiological protection system. Finally, remember that the criteria, regulations, and guidelines provided here are general, and specific requirements set up in national regulations are the ones that must be fulfilled for facility radiological compliance.

References

Alam S.M., Ansari R. and Athar Khan M. 2001. Application of radioisotopes and radiation in the field of agriculture: Review. *Journal of Biological Sciences,* 1:82–86.

Ashworth D., Shaw G., Butler A.P. and Ciciani L. 2003. Soil transport and plant uptake of radio-iodine from near surface groundwater. *Journal of Environmental Radioactivity,* 70:99–114.

Bateman, H. 1932. *Partial Differential Equations of Mathematical Physics.* Cambridge University Press.

Boniglia C., Onori S. and Sapora O. 2003. *Il trattamento di prodotti alimentari con radiazioni ionizzanti [Ionizing radiation treatment of foodstuffs].* Notiziario dell'Istituto Superiore di Sanità, v.16, Istituto Superiore di Sanità, Italy.

Brechignac F., Madoz-Escande C., Gonze M.A. and Schulte E.H. 2001. Controlled lysimetric simulation of accidents giving rise to radioactive pollution of the agricultural environment: Synthetic overview of research carried out at IPSN. *Radioprotection,* 36(3):277–302.

Comar, C.L. 1955. *Radioisotopes in Biology and Agriculture – Principle and Practice.* Oak Ridge Institute of Nuclear Sciences, McGraw-Hill Book Company.

De Hong L., Yang Y.-D. and Xiao C.-J. 2010. Multi-elemental analysis of food crops biological standard reference materials with instrumental neutron activation analysis. *Jiliang Xuebao/Acta Metrologica Sinica,* 31(5):33–36.

Delacroix D., Guerre J.P., Leblanc P. and Hickman C. 2002. Radionuclide and Radiation Protection Data Handbook. *Radiation Protection Dosimetry* v. 98, p. 168.

de Oliveira K.A.P., Menezes M.A.B.C., Jacomino V.M.F. and Von Sperling E. 2013. Use of nuclear technique in samples for agricultural purposes. *Engenharia Agrícola,* 33(1):46–54.

D'Sousa, S.F. 2014. Radiation technology in agriculture. *Journal of Crop and Weed*, 10(2):1–3.

Didi A., Dadouch A., Jai O., Tajmouati J. and El Bekkouri H. 2017. Neutron activation analysis: Modelling studies to improve the neutron flux of Americium–Beryllium source. *Nuclear Engineering and Technology*, 49(4)787–791.

European Commission (EU). 2000. *Practical Use of the Concepts of Clearance and Exemption – Part 1:* Guidance on General Clearance Levels for Practices, Radiation Protection, n. 122.

European Atomic Energy Commission (EURATOM). 1980. *European Council Directive 80/836/EURATOM*, Official Journal of the European Communities, n. L 246/1.

European Atomic Energy Commission (EURATOM). 1984. *European Council Directive 84/467/EURATOM*, Official Journal of the European Communities, n. L 265/4.

European Atomic Energy Commission (EURATOM). 1996. *European Council Directive 96/29/EURATOM*, Official Journal of the European Communities, n. L 159/1.

European Atomic Energy Commission (EURATOM). 2011. *European Council Directive 2011/70/EURATOM*, Official Journal of the European Communities, n. L 199/48.

European Atomic Energy Commission (EURATOM). 2013. *European Council Directive 2013/59/EURATOM*, Official Journal of the European Communities, n. L 13/1.

Holmes, R.J. 1982. Gamma ray and neutron sources (AAEC/S–24). In: Watt J.S. and Sowerby, B.D. (eds.). *Australian Atomic Energy Commission Research Establishment*, p. 123–136.

International Atomic Energy Agency (IAEA). 1963. *A Basic Toxicity Classification of Radionuclides*, IAEA Technical Report Series n. 15.

International Atomic Energy Agency (IAEA). 1973. *Radiation Protection Procedures*, Safety Series n. 38.

International Atomic Energy Agency (IAEA). 1996. *International Basic Safety Standards for Protection against Ionizing Radiation and for the Safety of Radiation Sources*, Safety Series n. 115.

International Atomic Energy Agency (IAEA). 1998. *Clearance of Materials Resulting From the use of Radionuclides in Medicine*, Industry and Research, IAEA-TECDOC-1000.

International Atomic Energy Agency (IAEA). 1999. *Assessment of Occupational Exposure Due to Intakes of Radionuclides*, Safety Guide n. RS G-1.2.

International Atomic Energy Agency (IAEA). 2000. *Management of Radioactive Waste from the use of Radionuclides in Medicine*, IAEA TECDOC-1183.

International Atomic Energy Agency (IAEA). 2001. *Use of Isotope and Radiation Methods in Soil and Water Management and Crop Nutrition – Manual*, Training Course Series n. 14, FAO/IAEA Agriculture and Biotechnology Laboratory and Joint FAO/IAEA Division of Nuclear Techniques in Food and Agriculture.

International Atomic Energy Agency (IAEA). 2004. *Application of the Concepts of Exclusion*, Exemption and Clearance. Safety Guide no. RS-G-1.7.

International Atomic Energy Agency (IAEA). 2006. *Safe Transport of Radioactive Material*, Training Course Series n. 1.

International Atomic Energy Agency (IAEA). 2008. *Security in the Transport of Radioactive Material*, Nuclear Security Series n. 9.
International Atomic Energy Agency (IAEA). 2009a. *Predisposal Management of Radioactive Waste*, IAEA Safety Standards Series n. GSR Part 5.
International Atomic Energy Agency (IAEA). 2009b. *Classification of Radioactive Waste*, General Safety Guide GSG-1.
International Atomic Energy Agency (IAEA). 2014. *Radiation Protection and Safety of Radiation Sources*, International Basic Safety Standards, n. GSR Part 3.
International Atomic Energy Agency (IAEA). 2018a. *Occupational Radiation Protection*, General Safety Guide n. GSG-7.
International Atomic Energy Agency (IAEA). 2018b. *Regulations for the Safe Transport of Radioactive Material*, Specific Safety Requirements n. SSR-6 (Rev. 1).
International Commission on Radiological Protection (ICRP). 1964. *Report of Committee V on the Handling and Disposal of Radioactive Materials in Hospitals and Medical Research Establishments*, ICRP Publication 5.
International Commission on Radiological Protection (ICRP). 1989. *Radiological Protection of the Worker in Medicine and Dentistry*, ICRP Publication 57.
International Commission on Radiological Protection (ICRP). 1990. *Recommendations of the International Commission on Radiological Protection*, ICRP Publication 60.
International Commission on Radiological Protection (ICRP). 1997. *Individual Monitoring for Internal Exposure of Workers* (Replacement of ICRP Publication 54), ICRP Publication 78.
International Commission on Radiological Protection (ICRP). 2007. *The 2007 Recommendations of the International Commission on Radiological Protection*. ICRP Publication 103. Ann. ICRP 37 (2-4).
International Commission on Radiation Units and Measurement (ICRU). 1998. *Fundamental Quantities and Units for Ionizing Radiation* – ICRU Report 60.
International Organization for Standardization (ISO). 1988. *Decontamination of radioactively contaminated surfaces* — Method for testing and assessing the ease of decontamination, ISO 8690:1988.
International Organization for Standardization (ISO). 2000. *Reference neutron radiations* — *Part 2:* Calibration fundamentals of radiation protection devices related to the basic quantities characterizing the radiation field, ISO 8529-2:2000.
International Organization for Standardization (ISO). 2016a. *Measurement of radioactivity* — *Measurement and evaluation of surface contamination* — *Part 1:* General principles, ISO 7503-1:2016.
International Organization for Standardization (ISO). 2016b. *Measurement of radioactivity - Measurement and evaluation of surface contamination* — *Part 2:* Test method using wipe-test samples, ISO 7503-2:2016.
Katz, L. and Penfold A.S. 1952. Range-energy relations for electrons and the determination of beta-ray end-point energies by absorption. *Reviews of Modern Physics*, 24(1):28–44.
Knoll, G.F. 2010. *Radiation Detection and Measurement*. 4th Edition. John Wiley and Sons Inc.
Lamm, C.G. 1979. *Application of Isotopes and Radiation in Agriculture*. IAEA Bulletin 21(2/3):29-35, International Atomic Energy Agency.
Laboratoire National Henri Becquerel (LNHB). 2020. http://www.nucleide.org/

Mendes, K.F., Martins B.A.B., Reis F.C., Dias A.C.R. and Tornisielo V.L. 2017. Methodologies to study the behavior of herbicides on plants and the soil using radioisotopes. *Planta Daninha*, 35: e017154232.

Milburn G.M., Ellis F.B. and Scott Russell R. 1959. The absorption of radioactive strontium by plants under field conditions in the United Kingdom. *Journal of Nuclear Energy Part A: Reactor Science*, 10:116–132.

Nandula, V.K. and Vencill W.K. 2015. Herbicide absorption and translocation in plants using radioisotopes. *Weed Science*, 63:140–151.

National Council on Radiation Protection and Measurements (NCRP). 2005a. *Radiation Protection for Particle Accelerator Facilities* - NCRP Report n. 144.

National Council on Radiation Protection and Measurements (NCRP). 2005b. *Structural Shielding Design for Medical X-Ray Imagining Facilities* - NCRP Report n. 147.

National Council on Radiation Protection and Measurements (NCRP). 2007. *Structural Shielding Design and Evaluation for Megavoltage X- and Gamma-ray Radiotherapy Facilities* - NCRP Report n. 151.

National Health and Medical Research Council (NHMRC). 1993. *Radionuclide Hazard Classification – The Basis and Development of a New List*. National Health and Medical Research Council and the Standards Association of Australia.

National Nuclear Data Center (NNDC). 2020. *Brookhaven National Laboratory*, https://www.nndc.bnl.gov/

Osborne, T.S. 2015. *Atoms in Agriculture – Application of Nuclear Science to Agriculture*, United States Atomic Energy Commission.

Particle and Heavy Ions Transport code System (PHITS). 2020. https://phits.jaea .go.jp/

Seprasert P., Anurakpongsatorn P., Laoharojanaphand S. and Busamongkol A. 2017. Instrumental neutron activation analysis to determine inorganic elements in paddy soil and rice and evaluate bioconcentration factors in rice. *Agriculture and Natural Resources*, 51:154–157.

Shalnov, A.V. 1976. *Some Activities of the IAEA on the Use of Radioisotopes and Radiation*, IAEA Bulletin 18(2):12-18, International Atomic Energy Agency.

Ente Nazionale di Unificazione (UNI). 1995. *Criteri per la costruzione di installazioni adibite alla manipolazione di sorgenti radioattive non sigillate [Criteria for the construction of practices for the manipulation of unsealed radioactive sources]*. UNI 10491:1995, Ente Nazionale di Unificazione, Milan, Italy.

Wheater H.S., Bell J.N.B., Butler A.P., Jackson B.M., Ciciani L., Ashworth D.J. and Shaw G.G. 2007. *Biosphere Implications of the Deep Disposal of Nuclear Waste - The Upwards Migration of Radionuclides in Vegetated Soils*. Series in Environmental Science and Management, Imperial College Press.

Index

Printed in the United States
By Bookmasters